Praise for *Astroquizzical*

'A wonderful jaunt through the universe at every scale, and a great way to fill in every gap in knowledge you have about astronomy.'
   Zach Weinersmith, co-author of the *New York Times* bestseller, *Soonish*

'*Astroquizzical* is a superb astronomy book, written with a distinctive tone which is both pragmatic and poetic at the same time. [Jillian Scudder] brings the perfect blend of fact and fascination to help us feel a greater sense of our place within the clockwork of the universe.'
   Jon Culshaw

'Scudder's mission is to provide the lay reader with a thorough grounding in the basics of astronomical knowledge. The writing is fluid and direct with the subject material brought vibrantly to life. For astro novices this book ... will bring a welcome depth to their appreciation of the night sky and the wonders it holds.'
   BBC Sky at Night magazine

'Genuinely entertaining [with] an excellent balance of enthusiasm and facts. This is the kind of book that would be excellent to get either a teenage reader or an adult with limited exposure to astronomy interested in the field. It's a cosmic journey that I enjoyed.'
   Brian Clegg, popularscience.co.uk

'The narrative form that Scudder employs is an imaginary cosmic journey that begins on our home planet and takes us in seven steps to the furthest galaxies. This simple format has been tried countless times before by big-name astronomers. What's different here is an intense level of engagement between writer and reader. Vivid storytelling explains the physics without equations. Her aim is to get people to think issues through for themselves, and that works. The clarity of Scudder's writing is impressive.'

Simon Mitton, *Times Higher Education*

'[This] excellent debut book is all about making complex concepts, if not exactly easy to understand, then at least a little easier to grasp. In her enthralling cosmic journey through space and time, astrophysicist Jillian Scudder discusses our home planet's place in the universe. Beyond the flawless presentation of known facts and current thinking, Scudder explores further by positing counterfactuals and thought experiments. The real triumph of Scudder's *Astroquizzical* is that it brings high-altitude, notionally abstract ideas to the general reader, presented in an entertaining and accessible way.'

Engineering & Technology magazine

'*Astroquizzical* approaches astronomy at a unique angle. It begins by stating that we are all distantly related to the stars; everything we're made of can be traced back to when they explode. By making this comparison at the start of the book, you instantly become intrigued and involved and from then on, the author – Jillian Scudder – does a fine job of covering a variety of topics and interests in space science.'

All About Space

The Milky Way Smells of Rum and Raspberries

The Milky Way Smells of Rum and Raspberries

... and Other Amazing Cosmic Facts

Dr JILLIAN SCUDDER

ICON

This edition published in the UK and USA in 2023
by Icon Books Ltd, Omnibus Business Centre,
39–41 North Road, London N7 9DP
email: info@iconbooks.com
www.iconbooks.com

First published in the UK and USA in 2022 by Icon Books Ltd

ISBN: 978-183773-101-5
eBook: 978-178578-927-4

British Library Cataloguing in Publication Data.
A catalogue record for this book is available from the British
Library.

Typeset in Carre Noir by Marie Doherty

Printed and bound in the UK

CONTENTS

ABOUT THE AUTHOR

Dr Jillian Scudder is an astrophysicist and associate professor of Physics & Astronomy at Oberlin College, Ohio, and the author of *Astroquizzical: A Beginner's Journey Through the Cosmos* (Icon, 2019). She has been writing about space for a general audience since 2013, with her work published in *Forbes*, *Quartz*, *The Independent*, and *The Conversation*, among others. When not teaching, speaking, or writing about space, she is an avid crafter, reader, and language-learner. She was born and raised in Florida, and moved to Minnesota, Victoria, British Columbia, and Brighton, UK (in that order) before landing in the Midwest and adopting an elderly cat.

PROLOGUE

I have often introduced myself to students on the first day of class by saying I have three main interests in life: space, dinosaurs, and volcanoes. I am delighted to let you know that this book has all three, though of course it's heavy on the 'space'. But then, my life is also heavy on the space bit; my profession is teaching physics and astronomy. Whether my class is current events in astronomy, a general overview course of outer space for non-science students, or upper level astrophysics, I really enjoy the process of teaching people what's out there, how we know it, and all the things we still don't understand.

Astronomers have learned a lot about the Universe over the years, and some of the things we've learned sound very sensible, like that there are planets around many of the stars in our own galaxy, and that our Sun is pretty average in most regards, and that a lot of things have to go just right in order to have a planet that can grow trees. And then there is a whole pile of things we've learned that just sound *silly*.

What kind of Universe comes with explosive moon volcanoes, planets as black as tar, and galaxies shaped like jellyfish, anyway? (Ours does.)

This book is a light-hearted tour through some of the more nonsensical things we know about outer space. As absurd as these may sound, everything here is extensively sourced to peer-reviewed publications; full details can be found at the back of the book, for those who want to know more, with a clear link back to let you know what part of the text they belong to.

I have done my best to summarize our current understandings of the topics discussed here. But, as with all science, these topics will be refined and re-examined over the years to come. Our interpretations of the observations may change. This doesn't mean that the observations we had before were *wrong*, but that we have more information now, and so the context for those measurements has altered. A great example of how dramatically interpretations can change in some circumstances: a planet we once thought might be diamond, we now think is probably lava. In other cases, our interpretations have remained stable, and – barring some unexpected new piece of information – are likely to stand the test of time.

In any case, as of the time of writing, this is the current state of affairs, and no matter what, this will stay true:

Space is weird, full of volcanoes, and it can kill you in many *very* creative ways.

THE ELECTROMAGNETIC SPECTRUM

	In low doses	In high doses	
Gamma rays	Radiation sickness	Death	 *High energy*
X-rays	See your bones!	Cancer	
Ultraviolet	Sunburns	Cancer	
Visible	Colorful!	Ow, too bright	
Infrared	Warm	Burns	
Microwave	Wi-Fi hotspot	Boils internal water	
Radio	Plays music	Music, but MORE	*Low energy*

THE UNIVERSE IS THE DIMMEST
IT'S BEEN IN BILLIONS OF YEARS

Given the number of stars in our night sky, it might come as a surprise to learn that the Universe hasn't been this dark in a long, long time – many billions of years.

Ten billion years ago, the galaxies in our Universe were forming a tremendous number of new stars – a number which has never been matched, before or since. When new stars are formed, they are produced in all sizes, from tiny, dim, red, and practically immortal, to the enormous, bright blue stars which live for approximately no time at all.* This difference in lifetime means that while you can see red stars for a long time, blue stars, due to their almost instantaneous deaths, flag a galaxy which is actively churning new stars into existence. And 10 billion years ago, there were more of these blue stars than at any other time in cosmic history.

To reflect this boom in star formation, this time period 10 billion years ago has been dubbed 'cosmic noon'. Which means that the time following cosmic noon can be (and sometimes has been) called 'cosmic afternoon', and so we can easily extrapolate that we will be proceeding onwards to 'cosmic

---

* They live for about 10 million years, but astronomically speaking that's a blink.

evening' and eventually 'cosmic night', which would presum- ably be when galaxies more or less stop forming new stars.*

If we take a census of all the galaxies we can find, and then order those galaxies by how far away they are from us (which is an easy way of putting them in age order, as we'll see below), the typical number of stars formed in a galaxy every year has done nothing but decline for the past 10 billion years. It's not a small drop, either – it's about a factor of ten. Meaning that every galaxy, typically, is forming ten times fewer stars than it was about 10 billion years ago. There are exceptions to every rule, of course, and there are still galaxies very near us forming ten times the number of stars that a 'normal' galaxy nowadays would produce, but there are star-forming overachievers in every era. Ten billion years ago, there were still galaxies form- ing an above-average number of stars; it's the average that's changed.

The exact graph which charts the decline and fall of the star formation rate of the Universe is called the Madau plot, after the lead author of the paper in which this accounting was first attempted, in 1996. To make this plot, we have to have three pieces of information. First, we need to know how far away the galaxies we're looking at are. This is an achievable task. Distance is a good proxy for how old these galaxies are, because the further away they are, the longer the light's taken to reach us, and so we're seeing them as they were, a longer time ago.

· ·

* This is different from the heat death of the Universe, which is when the expansion of the Universe is so fast that no stars *can* form; cosmic night might just be an exhaustion.

Second, we need to know that the bluer stars don't just produce *blue* visible light, they often also produce a lot of light that is so much on the blue end of things that the human eyeball is no longer able to see it – the ultraviolet. Ultraviolet light is most commonly referred to on packets of sunscreen as the thing we need to block; UVA and UVB are two chunks of ultraviolet light that unfortunately make it through our atmosphere. Most UV light is blocked by Earth's atmosphere, but the human skin surface has a bad habit of blistering when exposed to too much UV (this is what gives you a sunburn, a tan, or skin cancer, depending on exposure).

Third, we need to know how much ultraviolet light the galaxies are producing. This is partly easy – point a UV-sensitive telescope at the sky and see what comes in – and partly very tricky, because UV light is also easily blocked by many things, not just sunscreen. A layer of clothing, while not commonly found in interstellar space, is enough to block a lot of UV light. In a galaxy, what we have more commonly is Large Amounts Of Dust, and particularly so in galaxies which are forming a lot of stars. The very same galaxies which should be producing a lot of UV light, in fact, which is currently incredibly inconvenient for our goal of 'know how much UV light is being *produced*'.

Fortunately, there's a way around this, which is to observe the dust directly. If you shine UV light on dust, while the dust does a great job of blocking that light, the price it pays to do so is that it warms up. We're not talking *warm* warm – you couldn't make a slice of toast over it – but warmer than one would expect for a cloud of dust hanging out in the deepest voids of a galaxy. And, conveniently, this is also observable. You

3

just have to take a telescope in the infrared (too red for the human eye to see) and point it at the same galaxies you pointed your UV telescope at. The combination of the two – directly seen UV, and the heated dust clouds – lets you have an estimate of the *true* amount of UV being produced in that galaxy, and that's the data point you want to put on the Madau plot.

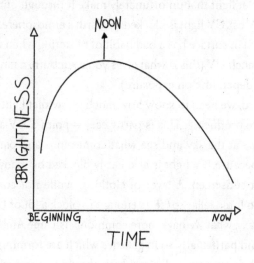

We've got 13.8 billion years of cosmic time to account for, and it seems that that first 4.8 billion years or so (cosmic morning) were also lower in stars compared to cosmic noon. This makes sense, because the *very* early Universe was just high-temperature soup, and to start building stars, you have to have gas that's cool enough to collapse down into a star. In High Temperature Soup Universe, all the particles are zooming around at high speed, and they won't settle down until the temperature chills out a little. The lower star formation in the earlier Universe tells us that the Universe didn't just immediately explode into a profusion of starlight as soon as everything cooled down enough; it turned the lights on a bit more gradually. But nonetheless, galaxies went relatively quickly from *dark* to the most star forming that there has ever been. The decline, since cosmic noon, has taken a much longer time than that initial rise.

The Universe is less blue now than it was 10 billion years ago. Those bright blue stars, formed in such abundance at cosmic noon, have all long exploded their fiery hearts out, leaving their placid redder cohort behind. Even stars like our own Sun only have lifetimes of about 8 billion years, so any stars that are even medium yellow will have burned out from that burst of star formation. If stars had completely stopped forming 10 billion years ago, we'd have a Universe full of very red, dim stars. Fortunately for our cosmic vistas, star formation didn't *stop*, it just slowed down, and so galaxies in our neck of the cosmic woods still have some blue stars kicking around; they've formed more recently, at a more sedate pace than in the past.

There's no particular reason to expect that this decline should pull itself back up — part of what made star formation so bright early on was that there was a *ton* of gas available to be turned into stars, and by now a lot of that gas is, well, turned into stars. So while the Universe is currently much dimmer than it was a billion years ago, it's also the brightest it will ever be, from here on out.

THE UNIVERSE IS BEIGE,
ON AVERAGE

If you're looking to assess the color of the Universe, it very much depends on the scale you're looking at. If you're looking at the inside of your house, the colors you find there will be very different than the colors in a deep forest, or on a glacier, unless your interior décor theme is Deep Forests And Glaciers. But if you're looking for the *average* color of the Universe, you can forget about glaciers and woods, firstly because as far as we know there is only one planet with forests,* and secondly because if you zoom out far enough, the only major sources of light are stars. (Though to be honest, they're very nearly the only sources of light, full stop. The colors of glaciers and the deep woods† are merely artistically reflected starlight.)

To ask what color the Universe is, on average, what we're really asking is: what color are the stars, typically, if we could view all of them at once?

In general, the stars in the Universe are collected into galaxies, and rather than trying to measure the colors of the individual stars in each galaxy, the much faster method is to look at the colors of the galaxies instead. All of a galaxy's light comes from stars, in any case, so it's not really cheating.

....

* It's ours.

† The deep sea, on the other hand, follows its own rules.

It is not, however, 100% straightforward. The light that we see from a galaxy isn't a pure average of the stars that are in it — other things exist in a galaxy which can modify its color, and the most dramatic of these is the presence of dust. Like sunsets on Earth, the presence of dust in a galaxy will redden the light as it passes through. It will also darken it, if there's enough dust. And if your galaxy's dust isn't evenly distributed (which it usually isn't), then you might wind up changing the color of some parts of your galaxy more than others.

So an easy first test is to try and get the average color of our own galaxy, where we have an extra high-definition viewpoint. The viewpoint is actually kind of a pain for this particular task; we are very literally struggling to see the forest for the trees. It's very easy to see the colors of individual stars in our galaxy, but finding the average color of *all* of them is much harder without an external perspective. Instead, we've tried to figure out what the average color of the Milky Way is by comparing our galaxy's mass and the rate at which it forms new stars (which are the main producer of blue light in the Universe) to other nearby galaxies, where we do have that external perspective, and assuming that ours lies somewhere in the middle of the color range for similar galaxies.

It's white.

As white as fresh snow, and it doesn't get much whiter than fresh snow on Earth. (Important note: this is an average color, and the Milky Way is a bluer color away from the center, and a redder color in the center, but these differences wash out to just plain white on average.)

7

If every galaxy were exactly like the Milky Way, the Universe should also be white, on average. If galaxies tend to be bluer than the Milky Way, then that would tell us that the Milky Way is forming less blue light than the cosmic average, which in turn means that it would be forming fewer new stars than the typical galaxy.

To find the average color of the Universe, what researchers actually did was find the average color of the *nearby* Universe. To find the color of the entire Universe, we'd have to be able to get the colors of *all* the galaxies, and that, without exaggeration, is impossible.

But we can take a census of all the galaxies which are relatively nearby, which gives us an average color of the current-day Universe. (We would have good reasons to expect that the average color should have changed over the last 13.8 billion years of the Universe's existence.)

Astronomers looked at one of the large, nearby galaxy surveys, the Two-Degree Field Galaxy Redshift Survey (usually abbreviated 2dF*), which gave them 166,000 galaxies to play with, in a bubble surrounding the Earth reaching out to 2.9 billion light years distant. If you can get the average color of the galaxies in this sweeping survey, you can get the average color of the Universe in this same space.

---

* You'll never guess how many degrees the Two-Degree Field survey covers.

It's worth noting that the researchers' primary goal here was *not* to calculate the hexadecimal color code[*] of the local universe.[†] Instead, they were using the color as a proxy for star formation (as indeed it is) to try to determine what the star formation history of nearby galaxies had been. The color of the Universe was quite literally a footnote on the third-to-last page of their paper.

It seems that, on average, galaxies in the nearby Universe are just a touch redder than the Milky Way, because the best descriptor for the color the human eye would perceive the typical galaxy's spectrum to be is 'light beige'. The human eye does a lot of averaging, with less sensitivity at the very red and very blue ends of the visible light range, so if we throw a galaxy's light at the eye, with all its mixture of reds, yellows and blues, the eye will describe this not in terms of the constituent colors but as 'white'. White sunlight, for instance, is an average of all the colors in a rainbow. The rainbow just gives us a means to see all of the colors spread out so that they are distinguishable individually.

The Universe being a bland paint color tells us a little bit about the history of stars in nearby galaxies. If the light from these galaxies was *very* red, we'd know that there haven't been

---

[*] Hexadecimal color codes are used mostly to tell websites what colors to display. They are a hash, followed by two alphanumeric digits for each of red, green, and blue. Each number ranges from 00 (none) to FF (all). #000000 is white, and #FFFFFF is black. A particularly pleasing deep blue can be identified by #0D0ACB.

[†] It's #FFF8E7, if you were curious.

very many young, blue stars formed recently, and that our galaxy is a bit unusual in being relatively bluer. And if the light from nearby galaxies was very *blue*, we'd know that our galaxy is forming an unusually *low* number of stars.

But instead, we get a color that's just a touch to the red side of pure white, which tells us that our Milky Way is not so far off the typical nearby galaxy after all. The Milky Way seems to be forming just a few more blue stars than the average, but not by a lot, and the light produced by a census of hundreds of thousands of nearby galaxies is close to white, overall. Our star, which produces white light, is thereby typical of both the Milky Way and of the many galactic neighbors we have.

The Universe near us, is, on average, a color not so far from indoor lighting. The researchers had one final challenge in front of them: trying to give a fancy name to a color that is just barely not pure white and a bit on the red-brown side. The authors said they were happy to call it anything '[a]s long as it's not beige'.* They took suggestions, holding an informal and non-binding referendum of nearby astronomers, and wound up picking 'cosmic latte' as their favorite, even though it didn't have the largest number of votes.† It was, however, caffeine-themed, and that was enough to override *vox populi*.‡

. . . .

* Oops?

† They didn't promise it was a democracy. The top voted (by Johns Hopkins University astronomers) was 'Cappuccino Cosmico', which admittedly has a lot more syllables.

‡ *Vox astronomi*, if you want to be picky about the Latin.

THE GALAXY IS FLATTER
THAN A CREDIT CARD

It's really hard to measure something when you live inside it *and* can't move around. And this is the problem we have with measuring the Milky Way. We've got the best resolution we could ask for, and we can examine this galaxy in far greater detail than any other galaxy in the Universe, but we are not blessed with any external perspective.

Instead, we are stuck on planet Earth, which moves only a little bit as it orbits the Sun, and so far from the center of the galaxy that it takes light, moving at 300 million meters every second, 26,000 years in order to travel between us and the center. It is not a centralized vantage point, nor a particularly special one in any other capacity.

However, our attempt to map out the galaxy is helped by the fact that it is mostly empty space, and the stars are relatively small, so we can in fact map out a good chunk of the Milky Way just by looking for stars, peeking through the gaps between other stars. This fails us in exceptionally dense regions of the galaxy (like the center) when the stars do actually manage to fill that region of the sky, but by and large it works all right. Well enough, in fact, that we've been able to build up a pretty good picture of the structure of the Milky Way by taking these maps and looking for patterns. Where are the stars most frequently found, and where are they rare? What kinds of stars are found in each area?

We've learned that the vast majority of the stars in our galaxy live in an incredibly thin plane – called the thin disk. This thin disk is also where our Sun lives, and so it's from the most populous part of the galaxy that we observe it in dark skies. If you've ever been fortunate enough to visit very dark skies, the Milky Way appears as a band of hazy light across the sky; this band is our galaxy, viewed edge-on and from within.

The fact that it *is* so thin in our skies lets us classify the Milky Way as one of many spiral galaxies in our Universe. Spirals are so named because, seen from above or below, their stars make up gently winding trails from the centers to their outer edges. Seen from the side, they're rapier-thin – and so the hazy light we see can tell us already that our home galaxy has spiral arms invisible to us from our place within it.

FROM ABOVE:

FROM THE SIDE:

Surrounding the thin disk is the thick disk. This is a less densely populated area of the galaxy, but there's enough here that we shouldn't ignore it. Different galaxies have different fractions of their stars in their respective thin and thick disks, but in general the thin disk always has more of a given spiral galaxy's stars.

What we're interested in here, fundamentally, is what the ratio of length to height is. Much like cakes can come as sheet cake (long, but not very tall) or in eight layers (not very long, but very tall), galaxies also come in a multitude of shapes. And to get a handle on how thin a galaxy is, all we need is a measure of how thick it is, and how large it is across, just as we might be able to measure the height and width of our cakes. Unfortunately, while you can measure cake with a ruler, for a galaxy, neither of these numbers is easily measured. And, in practice, all of our cakes are likely to be far too thick for this analogy to hold.

To compare a galaxy to plain letter paper instead, what we can do is take the surface area of a circle of paper, and the thickness of that paper, and compare it to the thickness of the stars in the galaxy and the size of the roughly circular galaxy.

Unfortunately, galaxies don't have hard edges. Galaxies don't so much 'end' as 'gradually fade out', and the longer you look, the fainter the edges you can capture, so at some point it's easier just to define some arbitrary threshold, like 'where the light is 80% fainter' or 'where 50% of the light is', and use that contour instead. Or* you could use the point at which the light

* Technically speaking this isn't an 'or'; astronomers use all three of these possible definitions in different situations. Half light radius is the 50%

has declined by a factor of 2.72,* and call that a 'scale length' or 'scale height' depending on which way you're going. The scale height is not a measure of the total height of the galaxy, but is an easier-to-measure *contour* of the galaxy, and it should work fine for estimating how much wider than tall your galaxy is. So let's start with the scale lengths.

The scale height of the thin disk of the Milky Way is about 400 light years, whereas the scale *length* of the galaxy is about 10,000 light years. The scale height of the thick disk is 1,000 light years, much larger than the thin disk, but then it only has 10% of the galaxy's light held within it. It's also not forming new stars, as it has almost no dust and no gas – all of that, which is required to make new stars, is held within the thin disk.

If we assume that the galaxy is a circle when seen from above (probably reasonable), and has a radius of 10,000 light years, and is 400 light years thick, we get a ratio of 25 : 1 (radius : thickness).

A piece of A4 paper's biggest circle will have a diameter of 21 cm (the width of the page), so that means a radius of 10.5 cm. And the thickness of letter paper, while it varies, is typically about 0.1 mm. A single sheet of paper is going to be thinner than the galaxy, if we keep these numbers, because 0.1 mm is 0.01 cm, which gives a ratio of 1,050:1. You'd need

----

contour, and another common option is the so-called 'Petrosian' radius, which is roughly an 80% contour.

* Formally, this is the mathematical number e (~2.71828), also known as Euler's number, and the base of a natural log.

42 sheets of paper to get the right thickness (4.2 mm) to match the galaxy. But these scale lengths aren't really tracing the full extent of the galaxy — just the inner portion where it is the brightest. To wit: this scale length of the galaxy is not even half the way out to the Sun, and we definitely, 100%, live inside the Milky Way.

By other measurements, the galaxy could be as much as 100,000 light years from center to edge, but this is an estimate attempting to place full boundaries on the galaxy, and not a characteristic 'scaling' that we might compare to other galaxies that we don't observe in such good detail. If we use the scale height of the thick disk, which should enclose almost all of the thin disk, then we're dealing with about 1,000 light years, as we saw above.

In these cases, we'd have a galactic ratio of 100:1. Still thicker than a sheet of paper, but you'd only need 10.5 sheets to catch up, making a pile only a millimeter high in total — just slightly thicker than a standard credit card. If you wanted to have a sheet cake hit the same ratio, with a thickness of about 2 inches (5 centimeters), you'd need a cake 33 feet (10 meters) across, which — to put this in perspective — if you're serving everyone a 2-inch by 3-inch slice of cake, would serve more than 5,200 people.

If you want to try pizza instead, which is maybe 3/8ths of an inch (9.5 mm) deep, we'd need one the size of a tall human being — 6 feet and 3 inches — across. This is obviously the sort of thing we can expect to find somewhere in the US, and indeed there is a restaurant in New York City that bakes a long, narrow pizza 6 feet long (about 1.8 meters) that alleges to feed ten people. (It's not even the largest in the US – the record for largest pizza you can buy is held by a restaurant in Texas which is 8 feet long and 2 feet wide.)

If we're going to have a small object that can fit in a standard kitchen, it'll have to be much thinner than a pizza. So let's try a credit card instead. Bank cards are a standard size: they're 5.4 cm tall and 8.6 cm wide and 0.75 mm thick. If you cut a circle out of one of these (not advised unless it's expired) you'd get a circle 5.4 cm in diameter (2.7 cm from center to edge) and 0.075 cm thick. This radius over the thickness has a ratio of 36:1, which is only a little off from the scale heights ratio we started with, and proportionally much thicker than the full extent of the galaxy.

The long way of the credit card gives you a ratio of 57:1, if you divide half the length by the thickness, which is still too thick to be proportionally right for the entire galaxy. However, you can get to about 100:1 if you stick two credit cards together longways. You'd need about six cards superglued together by their edges if you wanted to make a full circle out of the cards, which I especially don't advise, unless you've been hoarding a *lot* of expired cards.

It's rather impressive, to be honest, that we can determine a small household object that has the same proportions as the galaxy, all without leaving our houses. And if indeed we use our bank cards for scale, we can use them again to purchase a very large pizza with the same proportions, for ourselves and nine friends.

GALAXY COLLISIONS DON'T ACTUALLY CAUSE ANY STARS TO COLLIDE

There are lots of reasons to want to understand what happens when galaxies collide, and the least important of them is that they *look very cool*. The stars within a galaxy become dramatically stretched out, the galaxies tend to produce profuse numbers of stars, which in turn, churns up dust, and overall the galaxy can be transformed from a symmetrical, predictable system into a gravitationally improbable tangle. Within all that chaos, it seems unlikely that the stars within these galaxies should escape unscathed.

We don't need much fancy math to figure out whether they do or not – just a set of numbers we don't know very well. The first of those is how many stars the galaxy has, and the second is the general density of stars, but if you can't get that, you can make do with a general estimate of the size of the galaxy plus some assumptions we know are not quite right, but should be good enough.

To start with, we don't really know how many stars our own galaxy has – frequently the numbers 'somewhere between 100 and 400 billion' are tossed around – and we also don't know how big the galaxy is to any great precision, but we have a very fancy space telescope named Gaia that can do the density of stars. And so far it's given us a grand total of: 0.04 solar masses-worth of stars per cubic parsec, near the Sun. A parsec is 3.26 light years (i.e., the distance that light is able to travel in 3.26 years), which is just a bit short of the distance between the Sun and the next star over, or 30 trillion kilometers. It is a *long* way, and would make for a frankly horrific road trip.

This number of 0.04 solar masses is definitely hiding some things, though, because if we make boxes that are all one parsec to a side, this number is telling us that we should have 0.04 solar masses inside each cube. However, our Sun is 1 solar mass, by definition, which means to get to an average value of 0.04, to counteract the Sun's presence, we need to have a *lot* of completely empty boxes. To get this math down any finer, you have to start counting stars by how massive they are (which can be done), and at some point you have to figure out proportionally how often you get one really massive star instead of making a bunch of smaller ones, and that's another set of numbers that we don't know very well.

There's an old saying in computer science, which has been adopted by many other fields, that summarizes the idea that if you don't know your numbers very well going in, you also don't know the numbers coming out very well either: 'Garbage in, garbage out.'

But astronomers have never let this stop us, because if you're looking for a general sense of *how much garbage*, it doesn't particularly matter if you're off by a factor of four here and there. The end answer will be right to within a factor of ten.*

....

* The sheer number of times I've heard astronomers say: 'What's an order of magnitude between friends?' I have also heard physicists say: 'Two is basically one, so let's ignore it.'

So let's start with some rough numbers – it'll give us a general sense of how likely things are, and we can worry about the details later. What we're looking for is a likelihood that any given star in one galaxy will strike another star in another galaxy, given the volume of space it has to travel through, the density of stars, and how big those stars are.

Generously, let's assign 400 billion stars to the Milky Way. And then let's look at the companion galaxy that we will go smashing into: Andromeda, currently coming at us at 300 kilometers per second. Andromeda has about a trillion stars – it's roughly the same mass as the Milky Way,* and it's a bit larger, with a radius of 110,000 light years to the Milky Way's 100,000 or so.

If we decide that both galaxies are basically very flat cylinders (more or less correct), and that they are evenly populated with stars (definitely incorrect), we can get an estimate. This last assumption is a simplifying one, but not unreasonable, since the stars at the center of the galaxy will be more densely packed, and stars in the outskirts, and away from the plane of the disk, will be less densely packed, so it won't give an exactly accurate number, but it will tell us if it's *likely*.

So let's assume both galaxies are about 100 light years thick, and we have their respective diameters. This gives the Milky Way a total volume of 3.1 trillion cubic light years, and 3.8 trillion cubic light years of volume for Andromeda. With

---

* Not by a lot; some estimates put it only a factor of two different. Different mass estimates change which galaxy is more massive, but the difference is generally not very much.

400 billion stars in the Milky Way, this gives an average distance of one star per 78 cubic light years. This assumes each star is at the center of an empty box – to reach the next star over, we can find the length of the box, which in this case is 4.3 light years. We're not that far off from the distance to our nearest stellar neighbor Proxima Centauri, which is 4.25 light years away. The Andromeda galaxy works out to be one star per 38 cubic light years, as it has more stars in roughly the same volume, but we're only different by less than a factor of two,* which is about right given their general sizes. Any two stars in Andromeda would be separated by 3.4 light years.

Once we have a typical density of stars, we need to know the area that a star in one galaxy will sweep out in going through the other galaxy – this, in combination with the density of stars, will tell us how many stars it will typically encounter along its way.

Assuming that all stars are the size of the Sun,† we can find the cross-section of the Sun: the surface area that it presents as it moves through another galaxy. Basically, it's how big a target it is. An object the size of the Sun will be a circular target one Sun in size, and that circular target will then move through a series of other circular targets of all the stars in the other galaxy. If any of these targets ever overlap, you have a collision.

The Sun is 695,508 kilometers in radius. That works out to a circular area that's 1.8×10^{-14} square light years in size. It

....

* Astronomer-speak for 'close enough'.

† This is also wrong, but the Sun is more or less average-sized – probably a bit larger than is typical, if we're honest.

is a *very small number*. Even when you allow the star to travel through a vast galaxy, populated with stars 4.3 light years apart, that's *a very small number*.

A TYPICAL VOLUME IN A GALAXY

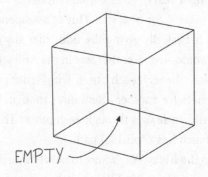

EMPTY

The shortest path one galaxy can take through another is if the two hit like stacks of pancakes that can move through each other. In that case, each star only has to go through the vertical thickness of the galaxy — in this case, 100 light years. The full path the star takes through the other galaxy sweeps out a total volume of 1.8×10^{-12} cubic light years, which we get by multiplying the circular target area by the length it travels (100 light years) to find out how much space that star ever occupied in the other galaxy. It's basically a cylindrical space we're tracing out, like shoving a straw into a fruitcake and wondering how likely it is that you've hit a raisin.

Now we have a volume of space, and we know the number of stars per volume, and so we can work out how many stars should live in that volume. (How many raisins should now be inside the straw?) If it's a large number, we'd expect stars to hit each other. But if it's a small number, much less than one, then we should expect that stars would easily miss each other.

A total volume of 1.8×10^{-12} cubic light years, multiplied by 0.013 stars per cubic light year (as above, based on one star per 78 cubic light years), gives a total number of encountered stars of 2.3×10^{-14} for one star. This is a stupendously low number, and is basically giving the odds that any one star in Andromeda would strike another star in the Milky Way, if they pass face-to-face through each other. They're not going to hit.

We can check for another orientation, though. What if the two galaxies slide sideways through each other? That's a much longer (and much less probable) path.

If a star in the Milky Way wants to travel through the entire length of Andromeda, that's 220,000 light years of path, which replaces our 100-light-year width. This gives a total volume swept out by a star of 3.96×10^{-9} cubic light years. Multiply this by 0.026 stars per cubic light year – from the density of stars in the more massive Andromeda galaxy – and you get 10^{-10} as your odds of encountering another star. This is still *a very small number*; they're not going to hit.

The spaces between stars are so vast, and the stars themselves are so tiny, that even a single collision between two stars would be exceedingly, outrageously unlikely. These sorts of numbers (one in a trillion odds) are also why we don't mind so much working with numbers that are out by a factor of four. Changing these numbers by a factor of four here and there won't change the end result – stars are tiny, compared to their galaxies, and the galaxies are mostly empty.

Even our assumption that stars are evenly distributed through a narrow disk isn't hurting us too much. In regions where the stars are less densely packed, the odds will be even lower. And in regions where the stars are more densely packed, they'll only be denser by a factor of a thousand or so,* which isn't enough to increase the odds of hitting another star significantly.

Beyond fun math, this exercise is sketching out an actual future event: the Milky Way is going to smash into the Andromeda galaxy, very, very slowly, over the course of the next 3–5 billion years. Our Sun, while it has a 0% chance of hitting another star, will nonetheless cause some problems on the same timescale; it's likely to become a red giant and onwards to a white dwarf around then. The charred husk of the Earth (if it isn't obliterated by the outer atmosphere of the Red Giant Sun) will travel, with the Sun, through whatever galaxy results from Andromeda and the Milky Way crashing through each other.

---

* Exceptions are made for the very center of both galaxies, where the densities of stars can be extremely high: 30 million per cubic light year, but only within the central 3 light years, so the odds of going there are very low.

THE GALACTIC CENTER TASTES OF RASPBERRIES AND SMELLS OF RUM

The center of our galaxy is not generally considered a place to go sniffing around; if you get too close to the exact gravitational center, you'll encounter the supermassive black hole's domain, which will comprehensively ruin your day. But if you stay a bit further out, there is a cloud of gas and dust, which has been studied extensively by astronomers.

If you'd like to point your arm at the center of the galaxy, you need to look in the direction of the constellation Sagittarius. You may recognize this constellation* as one of the members of the Zodiac, and its placement on the sky happens to coincide with the location of the supermassive black hole in the core of the galaxy. To name both the black hole and the cloud of gas which lives nearby, we affix the name of this constellation to them. The supermassive black hole at the center of the Milky Way is therefore known by the exquisitely appealing name Sagittarius A* (pronounced A-star).

The gas cloud nearby is also the Sagittarius cloud, which has been subdivided, equally appealingly, into Sagittarius A, B1, B2, C, D, and E. These sorts of names are the things a harried astronomer comes up with to label their data, and then they

---

* In the sky, its brightest stars look like a teapot!

stick because the astronomers who study the cloud all know which bit we mean, and no further thought is given to them. This particular gas cloud has roughly six largish clumps within it, and it's each of these clumps that gets a letter label.

It is in fair B2 where we lay our scene, though interesting science is happening throughout the A–E complex of gas. B2 itself is about 390 light years away from the central black hole, and contains a tremendous amount of gas, packed much more densely than a typical cloud of gas in a galaxy. This is nowhere near the density of gas on Earth, mind. The density of air can be described in terms of kilograms per cubic meter, while the inter-stellar medium – the very thin, dispersed gas between the stars in a galaxy – is usually measured in '*atoms* per cubic centimeter'. To convert the density of air into atoms per cubic centimeter, we can use Loschmidt's constant, which tells us there are 2.6×10^{19} (aka 10 quintillion, or 10 million trillion) atoms per cubic centimeter in air at standard temperatures and pressures on Earth.

The interstellar medium, is, on average, a heaping *one* atom per cubic centimeter, with a dense gas cloud clocking in at around about 100 atoms per cubic centimeter. B2, as an excep-tionally dense cloud, is around 3,000 atoms/cm^3.

It's extremely dense for a cloud of gas in space, but it's not what anyone would call 'breathable'.

The density of B2 (for a cloud of gas in space) lets it per-form more interesting chemistry tricks than the typical gas cloud, as the interior of the cloud of gas is a bit more protected from radiation which could split apart a molecule. This makes it a great place to go looking for which molecules *can* form in space, because they're less likely to be broken apart immediately after forming. The density helps us again, in that you might have enough of a given molecule to see its chemical signature from Earth.

All atoms, as well as the more complex molecules, have a particular set of colors they can produce given the right circumstances. The larger your atom, the more complex the set of colors produced, and if you have a molecule, these sets become even more complex yet. The complexity is both a boon and a burden; on the one hand, a molecule has a very unique fingerprint, but on the other hand, if you start to make a cloud of many different molecules and atoms, you can generate a forest of colors which can be difficult to disentangle.

B2's cloud of gas turns out to have the signature of a compound which, on Earth, helps give raspberries their flavor, and rum its taste: ethyl formate. And, leaning on research by other scientists, we know that this gas cloud doesn't just *smell* of alcohol, it also has *actual* alcohol in it. Ethyl alcohol has been found here too – this is also known as grain alcohol, or ethanol, and is the alcohol in beer and wine. It's also used in hand sanitiser and paint solvent, so proceed with caution.

We're well on our way to a Galactic Cocktail – if only you could concentrate the gas cloud down into a glass – tasting of raspberries and smelling like rum, with grain alcohol to spike it. It needs to be 10,000,000,000,000,000,000 times more concentrated to get to the density of water, but with enough dedication or time, that could probably be finagled somehow.

Regretfully, I must inform you that these tasty molecules aren't the only ones suspended in the cloud of gas. It has a few other companions detected so far, many of them poisonous:

formaldehyde (causes cancer), formic acid (will burn you), n-propyl cyanide (forms regular old toxic cyanide in the body) and methanol (toxic if you breathe, touch, or swallow it, and also catches fire), to name a few. Perhaps a cocktail for a murderous supervillain's cabinet, rather than your own.

So if we can't visit, sniff, or drink the gas cloud, why spend so much effort figuring out how toxic it is? What's the importance of sorting out exactly which chemical compounds exist in a gas cloud so distant from us?

We're trying to find out how easy it is to build DNA. Many of these studies are really looking for an amino acid; an early building block for DNA. If these molecules can exist naturally in regions where new stars are forming, it's a lot easier to build a planet out of the leftovers with amino acids either present, or within easy delivery range, after you've built a surface to stand on. If these complex chemicals aren't available, your would-be life-hosting planet has to start from scratch. Given the right conditions and materials, this latter pathway is certainly a viable one, but it takes longer than if your planet can give itself a jump-start. And, as the largest dinosaurs can attest, sometimes there are astrophysical setbacks to the development of complex life. An ill-placed asteroid – or even larger objects such as the Mars-sized object that created the Moon – applied to the crust of a planet is a dramatic event, but it does set back the development of complex life somewhat. Nothing says 'inhospitable' quite like a planetary-scale game of 'the floor is lava'. If you have amino acids lurking in comets and smaller rocks, you'd be able to start the culturing of life once more, a little more rapidly, once the ground is cool enough to stand on.

THE CENTERS
OF GALAXIES CAN BLOW
GALAXY-SIZED BUBBLES

If you're going to blow bubbles, you need two things: a source of wind, and some material that can deform to make your bubble. Typically we humans create the wind using our lungs, and the material that deforms is soapy water. Supermassive black holes, however, have invented a completely distinct, lung-free method of blowing bubbles.

Black holes, when trying to grow, are *profoundly* poor at actually growing. In order to grow, the black hole needs stuff to fall in through its own gravitational point of no return, but in order to lose that gravitational argument, the stuff, whatever it is, needs to either fall towards the black hole in a very specific manner, or it's going to have to slow down enough to fall in. The black hole can't really do anything in particular to help this process along, since it's just a gravitational lump, but it will serve as the end point for any gas that's falling 'down' within a galaxy. However, when the gas reaches the black hole itself, it will usually encounter *other* gas that got there first, all trying to fall into the black hole, but moving too rapidly to actually do so. The problem boils down to the same reason ice skaters are able to spin so fast; when they pull their arms in, they start to spin faster. Similarly, the closer to the black

28

hole, the faster the gas spins around, and so you wind up with an incredibly fast-moving, incredibly hot disk of gas around the black hole.

Some of this material will eventually fall into the black hole, but a lot of it will wind up dramatically ejected from the premises. Black holes have managed to build themselves a wind tunnel for electrons; truly profound amounts of energy can be flung out and away through narrow, funnel-like pathways in a narrow, finely pointed jet. Their formal name is the same: they're technically called 'jets'.

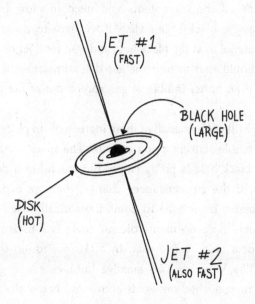

These jets will replace the need for lungs if you're trying to blow bubbles. But what about the material that can deform? Near a galaxy, that's not in short supply either, as most galaxies have a thin halo of warm to hot gas surrounding their disk. It's not universally present – if your galaxy is very, *very* small, then it's unlikely to have a hot halo of gas, and if it lives in a region of the Universe with thousands of other galaxies in the very near vicinity, then instead of having a personal hot halo, the entire cluster of galaxies has a collective, very dense halo of gas. But if the galaxy is massive (like the Milky Way), then there will be this reservoir of toasty, thinly spread gas that can be pushed around and deformed, if given an energetic nudge.

That's all the ingredients you need to blow bubbles, supermassive black hole-style. All we have to do is throw some material in at the black hole and wait for a jet to appear, and it should start to heat the gas that surrounds the galaxy into an even hotter bubble of gas, carved out of the galaxy's halo.

The Milky Way has all of these ingredients in place, minus the black hole actively trying to grow. The space around our galaxy's black hole is pretty empty, but much like a dormant volcano, if the circumstances change, then we expect the supermassive black hole to come dramatically back to life. And, much like a dormant volcano, there is substantial evidence for a more active past. In 2010, we found that our own Milky Way has two massive bubbles which glow in gamma rays, pushed outwards above and below the disk of the galaxy.

They're *enormous*. From top to bottom, they're about 50,000 light years in length. For scale, this is about twice the distance between the Sun and the center of the galaxy.

They're named the Fermi bubbles,* and are still an active subject of research. The Fermi telescope was only launched in 2008, so the bubbles' discovery came shortly after that, and required lots of careful data processing. These bubbles aren't incredibly bright, especially compared to the rest of the galaxy, which *is* very bright. In order to find them, the researchers had to be able to remove the brighter galaxy light, but once that was done, the bubbles were clearly visible. Lots of questions are now being asked, like: 'How long have those *been* there?'; and 'How *exactly* did the supermassive black hole *do* that?'

Some of these questions have been partly answered. It seems like they may have formed a few million years ago, which, in the lifetime of a galaxy, is *catastrophically recent*. The first paper put in an upper limit of 'sometime in the last ten million years', if they're expanding outwards at about 1,000 kilometers every second.

The second question – how *exactly* did the supermassive black hole *do* that? – is trickier to answer. We need some rather extreme source of energy, capable of inflating such large bubbles. The black hole is the most obvious culprit, since we see them elsewhere fairly regularly causing very bright jets to twang electrons entirely out of a galaxy. But the mechanism, and indeed the *when*, are both still taking suggestions.

One option is that the black hole in the center of our galaxy could have utterly destroyed a couple of stars. Or perhaps only

---

* Fermi is the name of the gamma ray telescope that took the data in which they were discovered, and which is in turn named after Enrico Fermi, an Italian-American physicist.

one rather large star. If you throw a 50-solar-mass star into the black hole and allow the black hole to shred it, you might be able to generate the energy required in 1,000–10,000 years. But you'd have to be pretty unlucky (or lucky, depending on your perspective) to send a star close enough to the black hole in order for it to be shredded this way.

Alternatively, the supermassive black hole might not have formed a narrow *jet* of material, but a broader, less focused wind, which we see in some other galaxies. In this model, the wind would have lasted for 10 million years, and ended only 200,000 years ago. This has the benefit of not requiring the extremely fast speeds of the jets, which cause some issues with observations of these bubbles.

If, on the other hand, it's an extremely fast jet that's responsible for inflating the bubbles after all, one model suggests that they may only be a million years old. Because this generates a bubble extremely quickly, the bubbles retain their incredibly crisp edges; if we left them alone for longer, we'd expect the edges to fuzz out.

As we keep hunting for the explanation to these bubbles, it's becoming more plausible that the supermassive black hole in the galaxy did indeed do something dramatic some 3.5 million years ago. It appears the black hole might have exploded so dramatically that it stripped electrons away from atoms in the Magellanic streams, which are at least 50,000 parsecs away – much further than the extent of the bubbles, which haven't gotten even a fifth that far. One particular model suggests that this explosion of activity could have also created the Fermi bubbles.

All of these possible explanations are incomplete, which is a point of Great Frustration — a recent review described them as having 'mysterious physical origin', which is the technical phrasing for 'we are annoyed by how little we understand this'.

Regardless of the exact mechanism that led to their existence, which will probably be debated for some years, the end result is firm: somehow, something at the center of the galaxy blew giant bubbles in the gas that surrounds the Milky Way.

A DISTANT BLACK HOLE IS
SURROUNDED BY WATER

There are times, in astronomy, when someone accidentally invents a whole new area of research just by trying to answer the question, 'What the heck is THAT?' The study of quasars is one of these. First detected in a radio survey as an unresolved, bright spot, which normally would mean you're looking at a star, these were way too bright to be a star, and so were dubbed 'star-*like*', and gradually the astronomical community settled on 'quasi-stellar object' or 'quasar' for short.

In the years since these early detections, we've finally worked out what the quasars are. They're supermassive black holes, at the centers of galaxies, very, *very* far away from us, and it seems the reason we're seeing them as such bright, tiny dots of light is that we happen to be looking directly down that extremely fast and bright jet of material that black holes throw out. It's always easier to see any light when it's pointed directly at you, which is why lighthouse lights spin, rather than just shine in one direction. So if the light from the black hole isn't pointed directly at us, it's going to be much harder to spot, but if it *is* pointed at us, we can wind up with an incredibly bright light shining directly into our telescope's cameras.

If you've ever gone out into bright sun after being in the shade for a while, there's an immediate 'Ow, my eyes' response, and it turns out that telescope detectors can do the same thing, but instead of yelling dramatically, what they do is bleed light from the pixel that should have detected it, into the neighboring pixels. This is called 'saturation', and when it occurs it's really hard to see anything else in the general area, because the camera is bleeding light sideways.

In order to be as bright as a quasar is, the supermassive black hole has to be trying very intensely to grow, and the best way to do that is to have a lot of gas in the disk around the black hole. Nowadays, many galaxies have exhausted whatever supply of gas they might have had in their very centers in their younger, star-forming party days. But if we go back in time far enough, there's a point when each galaxy is flush with gas, and so the regions surrounding a black hole can also be gas-rich, and easily able to fuel a quasar's bright jets.

Most of the gas that might be around a black hole is hydrogen, because there's a lot of it, but there's nothing requiring it to be purely elemental hydrogen, and molecules such as H_2O are also perfectly permitted. There's a particular quasar that holds the largest reservoir of water we've yet discovered in the Universe. Astronomers have done a Naming again, and this quasar is called APM 08279+5255. (It won't be on the test.) In this particular quasar, we're not looking at ice or liquid water; the water is a humid cloud of gas, and this is *only* the water that is just around the black hole, not a complete census of all the water in the galaxy. The water is warm by astrophysical standards, but it's still a pretty frozen −53 Celsius by human standards. In addition to being the *most* water, it is also the oldest glowing water we've yet found, dating back to when the Universe was only 1.5 billion years old. At the time of its discovery in 1998, it was the brightest object in the known universe.

This record-holding is no small feat – it's a *lot* of water: at least 25,000 times the mass of the Sun, entirely in water, though it won't be Galaxy's Finest Distilled Spring Water. It's mixed in with carbon monoxide and molecular hydrogen, probably among lots of other materials. NASA kindly summarized this mass as 'equivalent to 140 trillion times all the water in the world's ocean', which is even more difficult to imagine than 25,000 times the mass of the Sun. 25,000 times the mass of the Sun is so large that it's bigger than the mass of any known star in the Universe – the largest so far is 'only' about 200[*] times the mass of the Sun. The reservoir of water in this galaxy is equivalent to 125 of those stars. As another attempt at context, the entire Milky Way is thought to have 4,000 times less water in it, and most of that will be held as ice: this is genuinely a *very* large haze of water. The volume of water itself is impressive, but what's even more neat is *how* we found it.

This tremendous cloud of water molecules has effectively become a very low-frequency laser, shooting straight at us.[†] In order to make a laser (or a low-frequency one which gets called a *maser*), you only need a couple of things.

---

[*] The star R136a1 is estimated to have a mass of 215 times the mass of the Sun.

[†] *Pew pew.*

1. You need some cloud of gas, all made of the same material, and relatively dense. This could be neon, or it could be water, or anything else. In this particular case, it's water molecules.

2. You need some reason for all of that gas to be in an excited state – the electrons in the atoms need to have absorbed some extra energy, but not so much that they take off out of the atom entirely. In astrophysical contexts, to get the gas into this format means you need something bright and producing a lot of high-energy light (like X-rays) nearby. Since we're talking about a galaxy that has a very bright black hole, we've got that covered.

3. You need a little bit of incoming light, just to trigger one water molecule's electrons to dump some energy, also in the form of light. This is also pretty easily achieved; we're near a black hole, and even if that weren't enough, you have a whole galaxy.

What happens next is a cascade – the incoming light causes the water molecules to spit out light of a very particular frequency, which hits other water molecules and causes them to dump their own energy, which then hits *other* water molecules, and you chain-react your way through, creating a huge pile of light, all at the same frequency.

If the source of the energy – in this case the quasar – doesn't turn off, then you can re-energize all this water and keep your cascade going for a really long time. It's effectively a laser pointer that you can only see with a radio telescope, powered by a black hole instead of a battery. As long as the black hole keeps going, the water maser can keep going too; and this makes the signature of water in this galaxy much, *much* brighter than it otherwise has any right to be.

In order to spot this tremendous reservoir of water, we had to have a lot of things line up exactly right. If the black hole hadn't been so luminous, the water maser probably wouldn't have turned itself on. If the water weren't so dense and near to the black hole, equally, we probably wouldn't have seen it. And of course, if the jets of the black hole (and the water maser) weren't pointed straight at us, this would have been tremendously difficult to see at all. But because all these things did line up just so, we got to learn that water exists in really immense quantities, even when the Universe was only 1.5 billion years old.

SOME GALAXIES
LOOK LIKE JELLYFISH

In general, galaxies come in two shapes: pancake-shaped, and rugby-ball-shaped. Take a survey of the Universe, and almost exclusively, that's what you'll find.

But not entirely. Back in the early days of trying to do exactly this work, folks were dividing up the shapes of all the nearby galaxies, and they put all the pancake-shaped ones into 'spiral' galaxy boxes, and all the rugby-ball-shaped ones into a box labeled 'ellipticals', but there were a fair few that didn't match up, and so they wound up in their own box labeled 'irregular' or 'peculiar'. They were often asymmetrical, a chaotic mess of stars, gas and dust, and while it seemed they were forming lots of stars, like the spiral galaxies, they were often not very flat, but certainly didn't match the fuzzy smoothness of the elliptical galaxies either.

Since then, we've kept the classification of 'irregular' galaxies, but we've also gotten a lot better at understanding why exactly a galaxy is falling into that box. In some cases, it's a galaxy that is colliding with another, and the gravitational mass of the other galaxy is pulling the galaxy into unusual shapes. In all cases, though, the irregular galaxy is undergoing some unusual circumstances which are changing its appearance from one of the standard two classifications.

39

There are many ways to put a galaxy in odd circumstances, beyond just flinging another galaxy at it, and one of these other ways is how we get the delightfully named jellyfish galaxies.

Jellyfish galaxies started out as spiral galaxies, but while a normal spiral galaxy holds all of its gas and stars relatively tightly inside its disk, the jellyfish galaxies have tendrils of gas (and freshly formed stars) extending away from the disk, much like the tentacles of a jellyfish extend away from the bell.

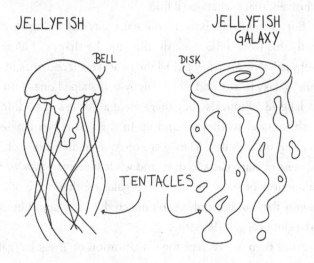

JELLYFISH

JELLYFISH GALAXY

BELL

DISK

TENTACLES

This is definitely an abnormal scenario for a galaxy. How did all of that material get flung out there? These tentacles are not a small distance from the galaxy – in some cases, this material, which presumably was inside the galaxy at some point, is tens to hundreds of thousands of light years away from the disk of the galaxy.

The key here is *where* we find these galaxies. They're *all* in large clusters of galaxies, and these patches of the Universe come with some extra features. For instance, the space between the thousands of galaxies that live in the cluster isn't empty. Instead, it's filled with a hot gas, which acts a bit like an atmosphere. And, like the Earth's atmosphere, it causes drag. This particular form of drag is known as ram pressure stripping, and it has the effect of acting like a wind, through which the galaxy must move.

If you or I were to step out into a strong wind, we might lose some poorly attached personal effects, but we wouldn't come *apart* from being breezed upon. The worst that can happen is something heavy, trundled along by the same wind, could strike us and cause an injury. But all my pieces will stay attached.

A galaxy does not have the same structural integrity. It's much more like holding handfuls of colorful powder in a breeze; the dust is easily scattered along with the wind. Galaxies, as they are mostly empty, cannot hold themselves together against any kind of outside force. And normally, there's no need; space is pretty empty, and so there's nothing to push against.

But in a galaxy cluster, we suddenly have a wind. Ram pressure is a general term for the force felt by anything moving through a fluid medium. Fluid here is loosely defined to include gas and liquid, both of which are capable of producing drag.*
We're familiar with this ram pressure sensation — any time

* Solid objects don't usually get to be moved *through* so much as *smashed into*, and that's not the kind of a drag we're looking at just now.

you've trailed your fingers through water, you're experiencing ram pressure. The pressure of the wind from an open window on an arm or face is the same thing again.

For a galaxy, this means that the pieces of the galaxy more prone to feeling this force (i.e., everything that isn't a star or otherwise a small dense object) will get blown away from the rest of the galaxy. This can result in the removal of a lot of gas and dust from the interior of the galaxy, and, like the streamers of dust, it casts them away in the direction of the wind.

Sometimes the breeze is gentle; this occurs when the cluster gas is thin, and it then can't exert much force on the gas and dust of a galaxy, and the galaxy will hold on to its material for longer. If, however, the cluster gas is thick, then moving through it at all will be very resistive, and the gas will be quickly removed from the galaxy into long streamers. And inside those rapidly removed streamers, we can form new stars, and that's when we start to be able to really easily spot these jellyfish galaxies.

The gas itself is visible to certain telescopes, like the Very Large Array in New Mexico, which can see cool gas directly, but once stars light up, these streamers very literally start to glow in the dark,* and it's much easier to see them.

The first identified jellyfish galaxies were found to be extremely bright, in fact. Follow-up searches, looking for more of what we now knew was an option, have found enough of them that we've been able to start doing more investigating into them. Beyond simply 'Whoa, look at this messed-up galaxy', we can now say 'Hey, these ones that are extra messed-up

---

* As do many actual jellyfish, under a black light.

are also doing these other strange things', and this line of questioning teaches us both about the cluster gas that's pulling these galaxies apart, and how a galaxy responds to this kind of treatment.

These galaxies are an active area of research; folks are still trying to understand how galaxies respond to being peeled apart like this. Their rarity tells us that clusters are not often so dense that the gas can be stripped so rapidly out of a galaxy; their luminance tells us that stars will form even in harsh conditions. Jellyfish galaxies should be in the process of changing from blue, gas-rich galaxies to redder spirals before too much (astrophysical) time* passes, since they are very rapidly losing the gas they require to form new stars. The star-forming tentacles will dissipate into the cluster, and the stars which formed within them will likely drift in intergalactic space until the ends of their own lifetimes. But what will happen to the spiral arms of the galaxy? How will the galaxy respond to the cluster environment, now that it has lost much of its gas? Are they extra rare because they only last for a short period? While we take the time to learn more about them, we will, much as we do with actual jellyfish, admire them from a safe distance.

---

* This is still going to take millions of years, so don't hold your breath.

THE WHOLE SKY GLOWS IN NEUTRAL HYDROGEN

Most of the Universe is hydrogen. This may seem odd for us humans, as we are made of lots of atoms that *aren't* hydrogen, but then we're also 70% water, and water is two-thirds hydrogen.

Any atom that's more complex than hydrogen has been constructed over the course of the lifetime of the Universe, in the core of a star or in a star's explosion at the end of its lifetime, so if we go back in time to the earlier Universe, we'd expect more and more of it to be hydrogen, the further back we go. And indeed, as we go exploring in the distant, younger Universe, we begin to find that many galaxies were surrounded by a cloud of neutral hydrogen gas.

This isn't unique to the distant universe – nearby galaxies also often have neutral hydrogen surrounding them, and rather a lot of it. In nearby galaxies, this hydrogen often extends beyond the bright stars within that galaxy, but if you look at it with the right wavelength of telescope, you can see it glow.

For nearby galaxies, the right wavelength is often the radio wavelength. Hydrogen gas, very rarely, will spontaneously spit out a little bit of light in this wavelength, as part of a random rearrangement process inside the atom. If hydrogen were rare in the Universe, this wouldn't add up to much. One tiny packet of light every 10 million years isn't anything to write home about, and if you were asleep, tough luck – you missed it.

But there's so *much* hydrogen that the sheer number of atoms outweighs the individual unlikeliness of the event, and so these thin clouds of hydrogen glow faintly, producing light with a wavelength of 21 cm, which is the same length as the short side of A4 paper. Large, earthbound radio telescopes can capture this light and map it out for us. And if we do this, we can capture the glow of the Milky Way first and foremost – just from our own galaxy, there is enough of hydrogen's shimmering faint glow that it fills the entire sky. This tells us that our galaxy still has the materials you'd need to keep forming stars for a while longer, and from it we learn a little more about how galaxies are built.

That's not the end of the glowing hydrogen story, though. If we look very far from us, to the light from galaxies which has been traveling the Universe for a long time, we can cover the sky *again*, with another layer of hydrogen. This time, rather than being in the radio wavelength, we need to shift to light that's much bluer than the eye can see.

The radio light that hydrogen can produce is a completely random process, but this very blue hydrogen light is formed through a different pathway. Here we need the hydrogen gas to have some nearby energy source – most easily, this is likely to be a young star or a supernova explosion. Both of these things are in abundance in the younger Universe (cosmic noon, after all), and so the early Universe is able to produce a lot of this extra glow.

Formally, this specific glow of hydrogen is called Lyman alpha, and it's in the ultraviolet. This light is produced when the lone electron around a hydrogen atom has absorbed some energetic light from its surroundings, and that electron has been left alone long enough for it to lose some energy again, which it does by emitting light at a very particular color. Unlike the radio light, which can be measured with sheets of paper, this light is measured in nanometers: 121.6 of them to be precise. For scale, a typical human hair might be around 60,000 nanometers in diameter, so this light wiggles up and down about 370 times in the span of a single hair.

Almost all of the light we can spot from the Universe gets a little stretched out as it travels, which means that the wavelength of light that's *produced* and the wavelength of light that we can actually *see* are slightly different – and the further away you are from Earth, the more this light is stretched. Longer wavelengths mean redder light, so this process of stretching the light and making it redder than it was when it was produced is called redshift.*

Sometimes this process is pretty convenient. In this particular case, the light as it is produced is too blue for the human eye, but there's a distance at which the light has been stretched out enough that it *does* fall into the boundaries of what we can see. Redshift is usually reported as a number, like 1.4 or 0.03 or 5. What this number really is is a fractional change – a redshift of 0.03 is a 3% change in the color of the light. A redshift of

. . . .

* Astronomers naming things, once more.

1.4 is a change of 140%. Redshift of 5 is 500% redder than it was when it was emitted.

When we're looking at the Universe as it was when it was only 2.1 billion years old, the light we receive is 300% redder than originally produced. With a 300% change, this Lyman alpha glow of illuminated hydrogen becomes visible. It's still too *faint* for the human eye to see, but that's why we invent telescopes which can capture the faintest hints of light.

If you keep going, and looking further and further, at some point you'll shift the light out the other side of what the human eye can detect – this occurs at a redshift of about 6 (600% redder than the original) when the Universe was less than a billion years old (0.9 billion), where the light fades out of the red side of our vision. But in between, in this range between redshifts 3 and 6, this glow of hydrogen from the distant universe is visible.

And if you go searching for it, it can be found. Researchers did precisely this, for a couple of very famous – but not *special* – patches of the night sky. The Hubble Deep Field and the Hubble Ultra Deep Field were both scanned, looking for hints of the glow of hydrogen. They found more than hints – in fact they found that this hydrogen *filled* the sky. And since the patches of sky they looked at aren't unique in any way, we should see roughly the same thing no matter which way we point our telescope, if minor details like 'the Milky Way is bright' and 'sometimes the Sun is in the way' didn't prevent us from looking in every direction.

Neutral hydrogen fills the sky twice; once in the radio wavelength, and once in the ultraviolet, neither of which are visible to the unaided eye, but which glow regardless.

What does this tell us?

First, it tells us that there is a *lot* of hydrogen gas swirling around galaxies in the distant Universe. We had kind of hoped to find the signs of this eventually, because one of the theories for how cosmic noon got so bright was that these galaxies have a lot of easy material to make stars. Much like how you can choose to make either a big fire, or one that lasts a long time, using lots of small sticks or larger logs, hydrogen is the fuel for galaxies to form stars – either rapidly, or for a long time.

This faint illumination from the past of our Universe has been invisible even to the largest telescopes for a long time, so to be able to finally capture this light, and confirm that yes, the presence of hydrogen is widespread in the early Universe, is a nice confirmation of what we had hoped to find. And perhaps we can also find some small delight in our continual bathing in light from the simplest atom in the Universe.

SOME OF THE STARS IN THE GALAXY ARE JUST PASSING THROUGH

Our galaxy is usually a pretty orderly spot. The stars almost entirely orbit in a flat plane, collectively swirling around the supermassive black hole in the center. Leave the galaxy alone for several million years, and we should expect to see everything pretty much exactly as it was. The stars in the galaxy are all on stable orbits, and while those orbits aren't usually perfectly circular, they're long, looping orbits, and the shape of their path is consistent.

There are exceptions to every rule, and so there's a population of stars which are doing something different. If you look carefully enough, you can find stars which are not just moving *much* faster than you might expect, but they're not even going in a sensible direction.

As far as astrophysical objects go, circular planes like the shape of the galaxy (or the solar system) are pretty simple. In order to maintain them, you typically have an arrangement where everything orbits some central mass, in the same general direction. There's very little motion in an up/down sense; while stars or planets may zip quickly *around* their central mass, they're doing so in a fairly narrow vertical space.

We also expect there to be a fairly consistent speed for things orbiting at roughly the same distance. The Sun, for instance, moves at some 220 km per second, relative to the center of the galaxy. The other nearby stars should be moving at about the same speed; relative to us, they're moving slowly.

Permit me to introduce you to the High Velocity Star. The high velocity star has as a defining feature (you'll never guess): extremely high velocity. It's what we can informally call 'fast'. They're uncommon in the galaxy, and one of the easiest and oldest ways to find them was to look for stars which were moving up (or down) away from the plane of the galaxy. This can be pretty tedious work; you have to be able to accurately measure a star's motion against distant background stars, and find the ones that stick out.

Fortunately, this is exactly the sort of thing a robot can do really well, and especially if you design the robot for this one particular task. The Gaia Space Telescope is exactly this. Its mission is to very, very, *very* precisely find the positions of a billion stars in the galaxy,* and then to watch those positions over the course of several years, which allows for the calculation of their speeds, relative to us. The speed relative to us can then be converted into a speed relative to the center of the galaxy, and then you can go on a hunt for the fastest stars.

---

* Fun fact: this is only about 1% of the stars in the galaxy.

There are three broad ways you can speed up a star. The first is to explode another star very nearby; the energy from the explosion can kick the star into moving much faster, the way a sudden large wave can carry an unsuspecting surfboard rapidly towards shore.

The second is via what's called a 'three body interaction' and this is mostly an effect when you have very compact assortments of stars, or near a supermassive black hole. When you have any three objects all strongly interacting with each other via gravity, the resulting paths they take are extremely convoluted, and often the least massive of the three winds up being flung out of the system entirely, at some extremely high speed. For a star to be sped up this way, it must come from the center of the galaxy, where the stars are packed closely enough together that this can happen, or it can interact with the supermassive black hole itself. We have found a few of these, which are dubbed High (or Hyper) Velocity Stars. Whether they're High or Hyper depends on how fast they're going.

The third way is if you have the same kind of three body interaction between three stars in a young stellar cluster, where these stars are just being formed. No black hole required. These stars, ejected from their formation cluster, are called Runaway Stars or Hyper Runaway Stars,* again depending on how fast they're going.

In general, these things wind up going faster than the average star in the galaxy, and in a different direction than we'd

* Astronomers: still uninventive namers.

expect, but unless they're going a thousand kilometers a second, however askew their stellar orbit is, they will continue to orbit the center of the galaxy.

This is where it starts to get fun, because if your fast star is fast enough, then it becomes what we call 'gravitationally unbound' – it's going so fast that the collective gravity of the entire galaxy isn't enough to hold it. It's leaving the whole galaxy, and is going to go wander intergalactic space for a while.* If the star is considered unbound, that's when the 'hyper' label begins to get attached.

Of 7 million stars in the current data release of Gaia, *twenty* were found to be candidates to be zooming out of the Milky Way. Seven of these twenty probably came from inside the galaxy – but not from interacting with the central black hole, since none of their orbits trace back to the center of the galaxy. Which means they were flung dramatically by a supernova explosion, or out of the center of some region of new star formation.

The other thirteen, however, are either falling *in* to the galaxy, or they're coming from some area of the Milky Way which has no real star formation happening – and you need that star formation to get a supernova, since the supernova happens at the end of a massive star's short life. When their orbits were modelled, they were unlikely to have crossed the disk of the Milky Way before.

. . . .

* Read: for the rest of its lifetime.

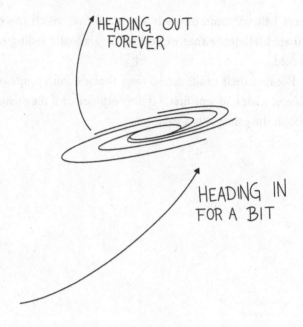

HEADING OUT
FOREVER

HEADING IN
FOR A BIT

Which means we've found thirteen interloping extragalactic stars. These were born in an entirely different galaxy, flung outwards by some explosion or strange interaction between stars, and journeyed through space, merrily shining their own light to illuminate their way, and happened to be observed by a robot we humans built. Since they're still shining, they can't have come far, and so the best guess is that they came from some disrupted dwarf galaxy that fell into our own, or from the Large Magellanic Cloud, another small galaxy which is in the early stages of falling into the Milky Way.

Because their orbits are so unusual, it's likely that many of these interlopers will only be temporarily in the galaxy. While some of them may be captured by the Milky Way's gravity, others will continue on their way until they reach the end of their lifetimes; either exploding or gradually fading into the cold.

Because their orbits are so *long*, though, this comparison between what happens first – if they explode or if they make it through the galaxy – is a valid one.

SUPERMASSIVE BLACK HOLES CAN SING A SUPER-LOW B FLAT

In the far-off realms of the Perseus cluster of galaxies, a super-massive black hole lives inside an equally impressive galaxy. The galaxy itself, called NGC 1275, is both the brightest galaxy in this cluster and host to some 2.43×10^{11} solar masses of material, and, as befits its status as the brightest galaxy, it sits on the gravitational throne at the very center of the cluster. Surrounded by a powerful magnetic field, this galaxy successfully siphons cold gas from its surroundings, guaranteeing itself a ready fuel line to feed the gigantic black hole that lives at its heart.

The inefficiency of this black hole's growth, like all other black holes, comes with the benefit of making its influence easy to spot. This particular black hole has been spewing energy outwards into the gas that surrounds the galaxy it lives within, carving enormous cavities and unusually shaped bubbles as a result.

Unlike the average galaxy, which lives a more isolated, socially distanced life, galaxies in clusters have thousands of galactic neighbors and are immersed in a thin haze of hot gas. This haze of gas is *so* hot, in fact, that it glows in X-rays. To glow at such high energies, this gas must be millions of degrees, earning it a solid 'do not touch' rating from Health and Safety. Conveniently, though, X-ray telescopes can observe this gas directly, as long as you're a moderately patient human. Studies of the Perseus cluster tend to count their observing time in hundreds of thousands of seconds (kiloseconds, to be technical about it). To translate into more typical units of time, a 200-kilosecond observing time (200,000 seconds) is about 55 hours* dedicated to staring at this one cloud of extremely warm gas, 240 million light years away from Earth.

What these hours and hours of observations with X-ray telescopes have been finding is that this ludicrously toasty cloud of gas is *even toastier* than it should be. If you leave a cloud of gas alone in space, you'd expect it to cool down eventually. And yet this gas is both tremendously hot, still, and it's full of weird bubbles. If it's going to be that hot, there must be something doing the heating. The black hole is the guilty party on both counts.

Two things seem to be happening in this cluster. One is that the black hole itself is blowing bubbles. We covered black hole bubbles a little earlier, but to blow a bubble really you just need some kind of pressure in some other medium, and in this case the black hole is inflating the already so-hot-it-glows-in-X-rays gas with even more high-temperature bubbles. These bubbles are very impressive and we're very proud of the black hole's artwork, but in this case they're hiding an even more fun trick, which is that the black hole is also humming a *very* low note.

* The longer you point your telescope at a patch of sky, the more light you can collect, so you'll catch fainter filaments. In technical terms, spending a lot of time on one point is called 'deep' imaging. The series of publications that investigated this cluster dubbed it 'Deep imaging' and then 'Very deep imaging'.

Any musical note is fundamentally a sound wave, and sound waves are themselves fundamentally a pressure wave. Generally speaking, you don't get to have sound in space, because there's no medium for a pressure wave to move *through*. However, in a cluster, we have this convenient extremely hot gas, which can serve the job.

Step one to uncovering this note is to remove the bubble from the image we received from the telescope. This is a slow and careful process, very specifically removing any hint of light that belongs to the high-temperature gas of the bubble itself.* When we do that, we're left with concentric rings of brightness, radiating outwards from the black hole. There are two things that could be physically changing in order to make the X-rays glow brighter and fainter in this way, and the first is that the glowing gas could just be getting hotter or cooler respectively. However, in this particular cluster, we know that the temperature isn't changing. The only other physical change we're aware of that can make the gas brighter is to compress it – changing the density instead. Since we have regular, concentric rings of changing density, what we have here is a pressure wave. Or a sound wave.

Sound waves are defined by their periods – how long does it take for the pressure to cycle from high to low and back again? The shorter the period of time, the higher-pitched the sound. The longer it takes, the lower the noise. The somewhat infamous 'teenager repellent' noise played outside some shops to discourage young people from hanging around is a very

---

* It's Photoshopping, but for science.

high-pitched noise that adults mostly cannot hear, but teen-agers certainly can, and find incredibly irritating.* (I can attest that this device usually gave me an instant headache.) This pitch, irritating as it is, has a frequency of 17.5 kilohertz, which means its pressure cycles from high to low and back once every 57 microseconds, and we can use this as a best-case scenario for the highest-pitched noise humans can expect to hear.

On the lower end, humans tend to be able to hear noises down to about 20 hertz, which cycles from high pressure to low pressure and back once every 0.05 seconds. This is a broad audible range, but it's nowhere near the range you'd need to have to process the sound created by the black hole in Perseus.

The oscillating pressure wave it creates cycles one time every 9.6 *million years*.

This is, weirdly, technically a note. It's 57 octaves below middle C – for comparison, a normal piano only has seven octaves. There was no guarantee that it would have wound up being an actual note; a similar phenomenon has been observed in the Virgo cluster, but the notes it produces, while even lower in tone than the Perseus cluster's sonorous bell, is decidedly off-key. Ranging between 56 and 59 octaves below middle C, none of the sounds it makes would be a pleasant tone, falling straight in the cracks in the metaphorical cosmic keyboard that goes down 60 octaves. (Don't try to build one – it would have to be more than 40 feet wide, just counting piano keys.)

* It won an Ig Nobel Prize in 2006.

SOME BLACK HOLES COULD BE NECROMANCERS

Nothing says we've entered the world of high fantasy like the entrance of a necromancer. With the power to raise the dead (usually to do their nefarious bidding), necromancers could not be further from the truth of a black hole, which is, after all, just a gravitational juggernaut. If we're going to call something a necromancer, though, there had better be at *least* something dead nearby to raise.

Good news: the stellar dead are available.*

Not all stars are like our Sun, which is both relatively isolated and free of stellar companions. Many stars are formed in pairs, and go through their entire lives orbiting each other. (For stars like the Sun, somewhere between 25% and 50% of them have a companion star.) If, however, these two stars are not the same mass, they will end their lives at different points in time. This is where we start to build fun stellar systems.

Stars which are more than eight times more massive than the Sun have a brief stellar lifetime in which they fuse hydrogen into helium, living about 10 million years (compared to the Sun's predicted 10 *billion*). They pass through all of their fuel at such a rapid pace that they will exhaust all their available hydrogen, become denser and hotter until they can fuse

* Terms and conditions may apply.

helium, exhaust all this, fuse carbon and oxygen, and end up as an elaborate elemental onion* of plasma. This plasma onion isn't stable or sustainable, and so these massive stars end their lives in spectacular fashion, undergoing a supernova event so cataclysmic that it flings the outer layers of the star away permanently, crushes the center of the star into a black hole or a neutron star, and for a period of a few weeks, can outshine the entire galaxy in which it lives. It's the Universe's best party popper.

Generally these events are described as happening in isolation, and often that's the case. However, if the massive star has a less massive companion, that less massive star wasn't done being a star yet – it was placidly fusing hydrogen into helium when its neighbor exploded. It's a good thing stars aren't capable of surprise, because if *my* neighbor suddenly became brighter than an entire galaxy, I give 100% odds I would startle.

What happens next depends strongly on how closely the two stars orbit. If they are far enough away, then nothing much occurs. The two objects – regular old star and black hole (or neutron star) – will continue to orbit each other. The crushed remainder of the star is now less massive than the star it was, since it exploded away a good chunk of its mass. If we let this stew for a few more billion years, we can come back after the less massive star reaches its own end of life.

Less massive stars, like our Sun, don't go through nearly as dramatic an end of life. They merely puff up to the size of our entire inner solar system, roughly reaching the orbit of Mars,

---

* Or a spherical parfait. (Everyone likes parfait.)

gradually lose gravitational track of their own outer layers, and leave behind a temporary* (though lovely) planetary nebula, and a white dwarf as a nugget of their former core, only about half as massive as the original star was.

Black holes, neutron stars, and white dwarfs are all considered stellar remnants. None of them are fusing anything, so they're no longer categorized as stars, but they are the ruins of what used to be a star. And, if you liken the fusion process of a star, with its creation of heat and light, to 'life', all three objects are now 'dead'. Great news if your career aspirations include 'be a necromancer'.

Let's leave neutron stars for now (more on them in a later section) and pass onwards to white dwarfs. A white dwarf is stable on its own, and we expect an unperturbed white dwarf to just hang around in the Universe for all of cosmic time,† gradually cooling from a white hot lump in the sky to a somewhat cooler lump in the sky. However, that's no fun, and there are lots of ways to make a white dwarf unstable. They all require a stellar companion, and depending on what the companion is, you can make your white dwarf periodically explode (a nova), completely self-destruct (another form of supernova‡), or, very specifically, if the companion is a black hole, you can temporarily make the white dwarf a fusion machine again.

. . . .

* They are visible only for about 20,000 years; a blip on a cosmic scale.

† In astronomer-speak, this translates to 'for all intents and purposes: for ever'.

‡ Type 1a supernovae are expected to be triggered by white dwarf objects self-detonating.

Enter the necromancy: no magic required here, because the black hole is going to make gravity do all its grave-digging.

The process involved is called 'tidal disruption', and is functionally the same as the way the Moon raises the tides on the Earth's oceans, but so strong that you can tear the object apart. Gravitational tides are a generic term for any time the force of gravity is substantially stronger on one side of an object, and substantially weaker on the opposite side. Ignoring complicating factors like 'rocks don't like to stretch' and 'gravity also holds the impacted object together', the impact of tidal forces is to stretch out the object, which our oceans on Earth obligingly do – stretching towards and away from the Moon.

If you are a black hole, however, you can turn this tidal stretching up to 11. Passing near, but not too near, to a black hole of the right mass can result in the stretching out of whatever object is passing through. It turns out that supermassive black holes are not ideal for this tidal stretching, because they're so massive that the white dwarf (or whatever else it is) will tend to fall straight in rather than stretching out first.

So let's say we've got the right mass of black hole, and we have a convenient white dwarf, about to pass near the black hole, close enough to feel strong tidal forces. How good is our necromantic resurrection to fusion? The key thing that needs to happen is that the temperature and density of the white dwarf have to increase. Simulations of this situation tell us that a white dwarf will get incredibly stretched along the direction that it's traveling, but incredibly *mashed* in the other direction. This compression in the 'up/down' sense is severe enough that you can trigger fusion where the white dwarf is most distorted.

'Temporarily' is a key point here, because it's worth emphasizing that black holes, while technically capable of becoming a necromancer, are not very delicate, nor are they selective, about how they apply gravitational force. If the white dwarf is close enough to the black hole to be raised back to fusion-life, it's also close enough to be absolutely shredded into oblivion shortly afterwards. What's not going to help the white dwarf stay intact is that fusion, in general, produces a lot of outward pressure. (In a stable star, this outward pressure from fusion is what keeps the star from collapsing inwards due to gravity.) In the simulations, this whole process takes about ten seconds; it's the closest, and final, piece of the white dwarf's orbit around the black hole.

Imagine trying to juice a tangerine by sending it through a pasta-maker, but where the rolls of the pasta-maker *also* create a huge wind on the way out. You've *technically* juiced the tangerine, but you have also sprayed that juice all over your kitchen.

The white dwarf is a mess after this encounter. Technically speaking it was raised from the dead and fused some more elements, but the price it paid was *all* of its structural integrity. What was a white dwarf is now a cloud of tidal debris, a completely disrupted ex-remnant which will not re-form. The white dwarf, though it may have had a brief foray back into the land of the stellar living, is doomed to return irretrievably to the land of the stellar dead.

The black hole *can* add 'Necromancer' to its business cards, as long as it doesn't mind counting ten seconds as a success story.

BLACK HOLE

STELLAR REMNANT
⇒ NECROMANCER ⇐

STELLARMASSBH@GRAVITYMAIL.COM
NO REPLIES POSSIBLE

Those ten seconds can be astrophysically useful, though – these simulations give us a prediction of what white dwarfs being torn apart should look like in the real Universe. So if we spot this, we'll know that we've managed to find the signature of a black hole that otherwise we wouldn't have been able to notice.

NEUTRON STARS COLLIDING GAVE US GOLD AND PLATINUM ON EARTH

When astronomers set about naming things, often we start with a reasonable name for a new phenomenon. In the case of a 'nova', which is a bright explosion produced by the leftover core of a medium-sized star, the term came from the Latin phrase *nova stella*, or 'new star'. These 'new stars' weren't so much newly formed as newly visible, as the remnants of a star will have sat around for many billions of years, but the sudden explosion of light they create can make them easily visible to us here on Earth.

And then of course we started observing *really* bright novae. So bright that they were clearly not quite the same thing, and to distinguish them, we gave them a prefix: supernova, to indicate that this was a high-test version of 'there is a brand new visible star here, from some kind of explosion'.

No problems so far, but then we found a population of novae that were too bright for a regular nova, but fainter than a supernova. Since they were about 1,000 times brighter than a nova, enter the 'kilonova'. And there were a couple of supernovae which were bright, even by the standards of a supernova, which had to be set apart and were given the name 'hypernovae'. So the astronomer's scale of prefixes, in order of brightness, currently looks like:

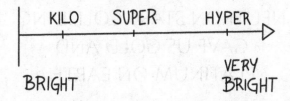

Up until a few years ago, kilonovae were mysterious events, vaguely postulated to be a collision between neutron stars. But, although these kinds of events are theoretically possible, a lot of things that are theoretically possible may not ever occur in the real Universe, either because the situations never arise, or just because they're so uncommon that a human lifetime is too short a viewing window to catch one.

However, on August 17th, 2017, the LIGO gravitational wave detector caught the signature of some massive objects slamming into each other, and the best model that fitted the data was of two colliding neutron stars, aided by the appearance of a bright explosion in a galaxy that matched the source direction of the gravitational waves. Pretty much every single telescope that could see the galaxy responsible had a look at it over the course of the next two weeks. Unlike a merger between two black holes, which — though violent in terms of space-time, and detectable to LIGO — is expected to be dark, a merger between two neutron stars is an incredibly luminous event.

Neutron stars are the stellar remnants of stars much larger than our Sun, but not quite large enough to form a black hole at the end of their supernova. The mass and density of these objects is so high that they are entirely made of neutrons — a very unusual substance indeed. They can come equipped with outlandishly strong magnetic fields, which then get called a 'magnetar' or a 'pulsar', depending on how we detect them.* To stand on them would be to get smashed flat; if a neutron star had the same mass as the Earth, gravity at the surface would feel about 700 million times more intense than Earth's. But in general, neutron stars are heftier, coming in at about 1.5 times the mass of the Sun, and so gravity at the surface is more like 100 billion times stronger than what we experience on Earth. Neutron stars are the land of the Extremely Flat Pancake.

And, like a white dwarf, we expect them, in the absence of anything dramatic happening, to be stable systems, just hanging around where their supernova left them.

Also like a white dwarf, we can make this more exciting by throwing things at it. A neutron star, like a white dwarf, can be found in binary systems with other neutron stars. These would be the remnants of what was once a system of binary high-mass stars, both of which have undergone a supernova. If you leave these stars alone long enough, some of them may gradually fall towards each other, eventually crashing into a single object – potentially, at that point, forming the black hole they weren't quite able to create on their own.† The other option would be to

----

* Pulsars are detected by the particular, lighthouse-like, flashes of radio light they beam outward from their poles. If we're at the right angle, we can detect this radio emission strobing past.

† Though, to be honest, even on the official LIGO diagram, the end point of these mergers is marked with a question mark, so this is a guess that may turn out to be outdated in a few years.

form a magnetar, a massive neutron star with an exceptionally powerful magnetic field.

Whatever the object turns out to be, the actual collision itself is a dramatic affair. It will be luminous across the entire electromagnetic spectrum, producing everything from gamma rays to X-rays, ultraviolet, optical light, and infrared light. The kilonova in 2017 was observed by some 70+ telescopes across nearly as many countries. After all these telescopes had had their fill of the light of this kilonova, it was concluded that these explosions were making an outstanding amount of the heavier elements in the periodic table. This was both a surprise and a relief.

A surprise, because the neutron star–neutron star collision had been, up until 2017, a purely hypothetical event, and yet here they were both existing and churning out tremendous amounts of the precious and radioactive elements we know on Earth.

And a relief, because we didn't have a good explanation for how those elements were formed, prior to 2017. And if kilonovae can do it, and they're relatively efficient at it, and not incredibly rare events (a few every million years or so counts), then let's dust our hands off and call it a day – that's a plausible explanation!

Plausible explanations are great, because if you'd like to have an explanation for nice things in your Universe, like animals, or cake, or your favorite human, or fancy jewelry in particular, the Universe has got a bit of work to do, because the elements that cake and your favorite human are made of don't come pre-built into it.

Most life as we know it is based on complex chains of carbon-based molecules, and jewelry is almost exclusively made of precious metals like gold, silver, or platinum. But when the Universe came to be,* it was made up exclusively of hydrogen, helium, and a tiny amount of lithium; the first three elements in the periodic table. The Universe's atoms are still mostly hydrogen (accounting for somewhere around 90% of all atoms), but clearly on Earth we are not made mostly of hydrogen, helium and lithium (though it's a great mixture to make a star), so some other process must have created the fancier elements you and I are made of, and also cake (if your cake is fancy enough you might put gold on it), and jewelry.

As anyone who wears jewelry can attest, there's no real shortage of gold, silver and platinum on Earth. They're precious metals, but not so rare that they're unobtainable. Granted, gold is usually found in volcanic areas where the heavy elements can be dredged back up to the surface, but with atomic numbers of 47, 79, and 78, silver, gold, and platinum are all well past the point we can get to with a regular star's fusion process, which stops at iron (atomic number of 26). So how is it that our backwater planet has so many of these metals?

You can make a small amount of some of these metals with low-mass stars in an unstable end-phase of their lives, but not very much. Supernovae of high-mass stars can abruptly create some heavier elements, but only really up to rubidium, with an atomic number of 37. But these elements are abundant in the

---

* However *that* happened, it certainly *did*.

chaotic, debris-filled mess that surrounds a collision between two neutron stars.

So our new solution to the presence of gold, silver and platinum here on Earth is that we must have had a neutron star explode somewhere in the general vicinity of the gas cloud that eventually formed our Sun and all the planets. If all these metals can be produced in a neutron star collision in relative abundance, then their relative abundance on Earth must be traceable to a historical neutron star collision.

In fact, someone's done the math on this. Given the abundance on Earth, and in the meteorites which trace the early solar system, they found that we really only needed *one* neutron star collision, which they model at about 300 parsecs (978 light years) away from Earth, about 80 million years before the formation of the solar system. That one explosion is enough to fill the Earth's reservoirs of heavy metals.

The radioactive elements curium and plutonium have short half lives, and so they decay quickly. But we see them preserved in specific quantities in meteorites which formed very soon after the solar system was condensing out of gas and becoming rocks, planets, and comets. It is possible to form these elements in supernovae, but supernovae are about a thousand times more common than kilonovae, and the model that best matched the data only had a single event.

The researchers found that 25–55% of all the plutonium, and 50–90% of the curium in the early system could have been flung our way by one neutron star merger. Using a metal with a longer half life (thorium's is about 14 billion years), they estimate that 30% of *all* the heavy metals in the solar system (which works out to a grand total of 10^{21} kg) were deposited by a single neutron star merger, about 5 billion years ago.

Statistically speaking, every piece of jewelry you own is one-third the cooled explosive barf from a *single pair* of neutron stars.

SOME OBJECTS SPIN SO FAST
THEY NEARLY SELF-DESTRUCT

Every so often, astronomers find the signature of something spinning around its own axis *unreasonably* quickly.

In general, the fastest of these are pulsars, a form of neutron star most easily found by catching the flashes of radio waves they send in our direction. Pulsars have had a long and glorious history of spinning uncomfortably fast: the first one discovered spun on its own axis once every 1.33 seconds, and that turned out to not be a particularly rapid spin, for a pulsar.

A lot of other pulsars spin more than once per second — an early catalogue of 96 pulsars identified a few that spin ten times every second. The range went all the way up to a lethargic once every three seconds.

These are already *very fast*. Pulsars are not tiny objects. I can very easily get a coin to spin once a second, but that's a coin. It's little. I can apply a lot of force to it, because I'm bigger than it is, and get it to spin pretty quickly for a while. But a pulsar is the remains of a *star*, and therefore has much more mass than the entire planet Earth.

Pulsars cannot, however, be *larger* than planet Earth, because they'd fling themselves apart.

Alright, so far so good. We can hold these objects together by making them very dense and small, in spite of their mass, and they might be able to spin up just because they used to be

larger, which is the same reason that ice skaters can spin them-selves up on ice by pulling their arms inwards to themselves.*
It's also the same reason that the Sun spins, because no one is flicking a blistering plasma ball to get it going, the way I can do with a coin.

As long as everyone's happy, let me introduce you to a new object that will comprehensively ruin the above framework.

Its name is B1937+21, and it is what we call a millisec-ond pulsar, because it spins once every 1.5 *milliseconds*. This means it's rotating 642 times *every second*. It was the first of this class of object, discovered in 1982, and remained the absolute record-holder for fastest spinning pulsar until 2006, when PSR J1748–2446ad came to steal the crown, spinning at 716 times every second.

I cannot spin a coin that fast.

In fact, this is much faster than even the fastest ice skater's spin – for which the record is an admittedly impressive 342 rev-olutions per minute,[†] which works out to 5.7 per second. The pulsars are *trouncing* the world's spinningest ice skater.

Millisecond pulsars cause all sorts of existential trouble. First of all, how did they speed up that fast? These two mil-lisecond pulsars have surfaces that are moving at speeds so fast that writing them as fractions of the speed of light makes

---

* Conservation of angular momentum! Angular momentum is a combin-ation of rotation speed and how far away the mass is held from the rotation axis. If you have mass far out and slow rotation, when you bring it inwards you have to speed up the rotation to keep everything in balance.

† By Olivia Oliver, in a charity drive.

sense: the surface of the older of the two is moving at 13% the speed of light.

And secondly, how are they not coming apart at the seams? There's a balance that must happen here between the force of gravity, which pulls inwards, and the centrifugal force felt by a rotating object, which goes outwards. There's a rotation speed at which the centrifugal force should win, and the object will shred its outer layers off and outwards, no longer able to hold them in by even a tremendous gravitational force.

If these objects are still here to be seen, they must still be holding together, which means that they must be small, even though they're spinning so rapidly.

We can play this game for the Earth. Let's say we want to spin up the Earth until it comes apart.* This ignores the inconvenient fact that rock likes to stay in one piece, and is just playing the game of 'How fast does the Earth spin before I'm no longer held on the ground?'

We know the gravitational force of the Earth, so we just need to find the speed at which the centrifugal force balances it, and we know it's going to be *quite fast* and much faster than the current spin rate.

It works out to just a touch slower than one rotation every two hours. At a two-hour day, the Earth would start to lose its outer layers and any other objects not nailed to the ground, as the speed of rotation would start to balance the gravitational inwards force.

. . . .

* Let's say you're a supervillain.

If you've ever been on one of those playground round-abouts* where you (or a friend) can spin it up, there's a point where your ability to hold on to the bars is exceeded by the speed at which the thing is spinning, and you will slide with varying amounts of grace off the merry-go-round. This is exactly the same effect that we're dealing with, on a much larger scale, when we talk about objects spinning too fast in space.

The Earth breaks up before it starts spinning anywhere near the speed of a pulsar, because it's 1) much less massive than a pulsar and 2) much *bigger* than a pulsar, and both of these contribute to the balancing point rotation speed.

A pulsar is often only 10 km or so from center to edge (compared to the Earth's 6,371 km for the same distance), and the pulsar has smashed around 2×10^{30} kg (one entire Sun, and about a million times more mass than the Earth) in that much smaller space.

The newly crowned Fastest Pulsar, PSR J1748−2446ad, must be less than 16 km from center to edge − and we know this because if it were any larger than that, it would be tearing itself apart.

All well and good − we have objects so dense in the Universe that they can spin 700 times a second without shredding themselves into oblivion. But if the typical pulsar only spins once a second or thereabouts, we still have a very excellent other question: How do you make a pulsar spin that fast?

----

* These are now considered 'dangerous', for pretty much exactly these reasons.

For the most part, these extremely speedy pulsars all have a common feature: they have a companion star. This other star seems to be the critical component to getting these pulsars sped up as far as they are. The pulsars are effectively cheating off their neighbors. At some time in the past, these neutron stars would have had a massive star as a companion, and the neutron star would have been able to peel off the outer layers of that companion star and wrap them around itself. This process had the effect of having a friend on the outside of your roundabout, giving the thing a kick every few seconds, gradually speeding up the rotation more and more.

Often, by the time we're looking at these pulsars, those companion stars have also died and left behind only a small remnant that marks their former location. But it's enough to show us that there *was* a star there, and when we spot these ex-stars next to so many breakneck pulsars, we can figure out how they might have gotten so dramatic. It's by stealing from their neighbors.

IT RAINS IRON ON SOME
BROWN DWARFS

Filed under 'How *Do* Stars Work?' is the comedically extreme weather encountered outside of our solar system. Example: it seems that there is liquid iron raining out of the clouds on some failed stars.

Brown dwarfs, before we get to their weather patterns, are already strange objects in our Universe. They're massive lumps of gas that are too large to normally be considered a planet, but aren't big enough to be a star, either, which has earned them the inglorious title of 'failed stars'. They don't have the mass to fuse new elements to produce heat and light the way our Sun can, but rather are mostly warm simply because of their own collapse down from a diffuse cloud of gas into a strange-behaving lump. Since they're not generating very much new heat (they can form a little, for a short while), they spend most of their lives being slowly defeated by the cold void of space.

They're not very warm, in the end. Most of them are some-where in the range between a 'No thank you, too hot for me' 1,000 Kelvin* down to a more reasonable (though still liter-ally blistering) oven or cup of coffee temperature. (One of the coldest is below the temperature at which water would freeze.)

. . . .

* To get from Kelvin to Celsius, add 273 degrees.

They're much warmer down in their cores, but even there we know their temperatures have to stay below the admittedly summery 13 million degrees Kelvin required to fuse elements.

If you are a star, or even a failed star, the amount of light churned out at the surface is directly related to how hot that surface is, and so because these failed stars are only 'meh' levels of warm, they're correspondingly not very bright. They're most easily seen by infrared observatories which search for the glow of warm things, but even then, a brown dwarf is still pretty dim. They're *so* faint, in fact, that we had to wait until 1995 for our technology to get sensitive enough to be able to detect them.

Another side effect of brown dwarfs not being very warm is that you can get interesting things happening in their atmospheres. A normal run-of-the-mill star is so warm at the surface that everything has gone one step beyond 'being a very warm gas' and has gone all the way into being a plasma, which is what happens when you set the temperature dial of a gas to 'Absurd'. It doesn't matter what the element is, or how much of a solid object it might be on Earth. For instance, iron, on the Earth, is a very convenient ore, and decidedly a metal, solid object. On the Sun, it's been stripped of many of its electrons, and is present as a plasma. Even at the surface of the Sun, where the thermostat is set as low as it's going to get (5,600 Celsius) iron is still missing a lot of its electrons and absolutely still a gas – nowhere close to being a lump of solid metal.

On a brown dwarf, you can start to get clouds. This is behavior we expect, because if we compare the Sun to, say,

Jupiter, which is our biggest planet,* the Sun's atmosphere is still extremely hot and made of plasma, whereas Jupiter's atmosphere is thick with opaque clouds and complex storms. In bridging the gap between the Jupiters of the Universe and the stars of the Universe, the brown dwarf population has kept the clouds and storms of Jupiter, and made them its own.

Some studies have tried to map out the surface layers of two nearby brown dwarfs. They found that their surfaces vary considerably in brightness with time, and not only that, the longer they watched, the more variability they found. This, the authors decided, was evidence for clouds.

It's a sensible conclusion; if the surface of a brown dwarf is changing in brightness, that means that some patches are brighter or darker than others. If you're looking in the infrared, then this translates into hot and cool patches on the surface. And an easy way to get hot and cool patches is to have cooler clouds obscuring the warmer depths, in exactly the same way that the clouds do on Jupiter.

Pair this with earlier work studying how the gases present in our Sun would start to behave at cooler temperatures, and we do expect to find layers of clouds on the cooler brown dwarfs. Not only that, but we'd expect elements that are typically metal on Earth (like the aforementioned iron) would be carried towards the surface of the brown dwarf like bubbles in boiling water. At some point, the iron would to start to condense out of the gas state it's been in, and become liquid as it

* Aside from one of them being blindingly bright and the other one not doing that (minor details).

cools at the surface. Once you have a gas condensing into a liquid and falling back down – hey presto, you have got yourself some rain. It's just that it's liquid iron, instead of water.

Oh, and also sand.

Because what's worse than iron raining from the sky? *Sandy* iron raining from the sky, at more than a thousand degrees.

The conditions under which iron is a liquid are conditions I do not like to experience myself. On Earth, these are met primarily inside blacksmiths' forges or in the equally furnace-like depths of a volcano, neither of which are great places to hang out, even if you want a fully cooked pizza in seconds flat.

In fact, the more brown dwarfs we look at, the more common the variability in their brightness seems to be, which we take as the signature of massive storms. Half of another sample of brown dwarfs showed this same kind of change with time. If you account for the fact that we don't get to choose the viewing angle of the brown dwarf (if we're looking at them from the 'top' down – the top being one of the rotational poles – they won't change so much, because instead of seeing clouds enter and exit the half of the brown dwarf we can see, they'll just spin around the pole), it may be that 'most' or 'all' of the brown dwarfs have these tremendous storms.

Sometimes these storms seem to blow over in less than a day, while others may linger, but we're limited by how much time we have on the telescope we want to use, and lots of people also want to use it for equally interesting science, so it's hard to say. You'd have to continuously monitor for a long time to get information about whether these brown dwarfs have storms like Jupiter's Great Red Spot, which has lasted centuries. (Though since Jupiter can maintain one, and it's probably not special,* some brown dwarfs are probably likely to do the same thing.)

Some of the temperature variability seen in these brown dwarfs might be us watching bands of material rotate into and out of view. This is a lot like the bands of material seen on Neptune, except instead of being made of methane and ammonia ice, it's clouds of iron, silicate and salt. At least for one brown dwarf, the clouds are more stripes and less spots. But this is still pretty consistent with the gas and ice giants present in our solar system; Neptune has a few spots, but the bands (less visible to the naked eye than they are on Jupiter) are what drive Neptune's catastrophically strong winds.

Oh yeah. Did we forget about the winds? The sand and iron raining down from the sky is *blasting* along through screamingly fast winds, sometimes up to 1,400 km per second. There are places that we should not go visit, and Sandblasting Brown Dwarfs are probably high on the list,† unless you want to look like the bad guys at the end of *Raiders of the Lost Ark* in 0.2 seconds.

---

* This kind of argument, the 'Well it's happened once, and we're probably not special', is a very particular flavor of 'The Universe is big, and weird things happen' line of logic. If it's happened once, it's likely to happen again somewhere.

† Don't worry; the list is extensive. Space is full of places that humans could not survive in.

SEVERE WEATHER
⚠ ALERT ⚠

================================

NIGHTMARISH WEATHER
TO CONTINUE FOR THE NEXT
BILLION YEARS

In case you thought we could just avoid the iron rain by avoiding brown dwarfs, bad news: brown dwarfs aren't the only place you can get iron rain in the forecast. The critical piece is to have iron in the atmosphere, and heat it enough to vaporize it, and then let it cool down somehow. For a brown dwarf, the cooling is done by getting far enough away from the heat source, which is the center of the brown dwarf.

If you're a massive gas giant planet you can do the same things, but the 'get away from the heat' maneuver is now rotating to the night side of the planet. We've seen the conditions for this to happen met on at least one planet outside our solar system – on WASP-76b in particular. As the planet's atmosphere turns from day to night, at that border, the air cools enough and the iron could rain out. WASP-76b has the same kinds of temperatures as a hot brown dwarf: the day side hangs out around 2,700 Kelvin, which is enough to vaporize the iron in the atmosphere, but the night side is only 1,800 K, nearly a thousand degrees cooler. This temperature drop is enough to permit iron to condense out into rain.*

It'd make for a dramatic sunset.

----

* This particular exoplanet is still under active study, and there are other explanations for the disappearance of iron from our measurements of the atmosphere, aside from rain – like the formation of clouds of aluminum oxide. In a crystal form, aluminum oxide is the stuff of rubies and sapphires.

WE SAW A CHUNK OF ROCK
OR ICE FROM OUTSIDE
THE SOLAR SYSTEM

To date, the solar system has had two visitors from a great distance. The first was detected in 2017, and given the name 'Oumuamua, in consultation with Ka'iu Kimura and Larry Kimura.* This name means scout, or messenger, and 'reflects the way that this object is like a scout or messenger sent from the distant past to reach out to us'. Together with the second interstellar visitor, named Borisov after its discoverer, we've now made a complete list of everything we know, so far, that has swung through our solar system from outside it: a grand total of two. There have almost certainly been others we haven't seen, and there will probably be more, though they will be rare — and impossible to predict.

One of the downsides of trying to learn about small, fast-moving balls of rock and/or ice as they cruise through the solar system is that because they're small and very fast-moving, once you find one, you have *very* limited time to learn as much as possible about it before it gets too far away and too faint to see, even with powerful telescopes.

....

* 'Oumuamua was discovered by a telescope on Haleakala, Hawaii, and so the name is in Hawaiian to help honor the indigenous culture, on whose traditional lands the telescope was built.

By the time 'Oumuamua was discovered, more than half of the time it would ever spend inside our solar system had already elapsed. It had entered from above, zipped underneath the Sun and emerged above the plane of the solar system a little bit beyond the location of Earth – this is where it was when it was spotted. It was already on its way out again.

We had about two weeks to observe 'Oumuamua before it faded from sight. We learned that it was an odd shape, long and narrow. It was likely ten times longer than it was wide, at only about 80 meters across, but some 800 meters on its long axis. It was also tumbling chaotically as it moved through our solar system, end over middle. We could tell it was tumbling, because the amount of reflected light changed dramatically with time – something easiest to explain if you change the surface area that can reflect light back at the Earth.

From that reflected light, we also got a handle on what it was made of, and that seemed to line it up closest with comets or icy asteroids within our solar system. But strangely, in every image it was a solid dot, without any haze surrounding it. It seemed to be behaving like a rocky asteroid, and not like a comet, which was what we'd expected. It's easy to throw comets out of a solar system; a large number of them hang out in what's called the Oort cloud, hugely far from the Sun, and they are readily dislodged.

We immediately have a lot of questions. Why does it have such a strange shape? This is more elongated than anything we know of in our solar system. Why isn't it behaving more like a comet, if it is made of lots of ice? Where did it come from? How long has it been traveling the galaxy?

There was, of course, one more mystery to throw into the mix, just for good measure. As 'Oumuamua was leaving the solar system, it started to go just a little bit faster than we expected if it were just a rock. It wasn't being guided purely by gravity, in other words. But there was an easy explanation for this, because we've seen it many times before with comets. Comets can eject gas from their centers and effectively rocket-propel themselves forward and give themselves a little bit of extra speed, and this is what was postulated for 'Oumuamua. It was a little bit puzzling, since 'Oumuamua hadn't shown obvious signs of a big cloud of gas around it when it was closer in to the Sun (and therefore warmer), but you don't need much gas to vent to explain the observations, and that much gas might have been invisible to the telescopes observing earlier. It should have vented some dust too, which would have been visible, unless of course it was larger dust* than is found in most solar system comets, which we do see occasionally.

The best guess was that it was fundamentally a tiny comet, but that it had a little bit of a protective coating; the extremely undelicious equivalent of an ice cream bar. Ice on the inside, some other not-ice on the outside. And so, as it left the solar system, our ice cream bar interloper had sprung a leak through its outer crust, and this leak helped it along in its escape. It was likely just doing some Comet Tricks and venting gas in a way that any interplanetary spacecraft propulsion system would recognize.

....

* Tiny dust particles scatter visible light better than larger particles, which makes the smaller ones easier to see.

Alas, there was a window of time during which news outlets did us all a disservice and went with the always incorrect theorem that this slight speed-up of 'Oumuamua might be evidence that, instead of a cometary-type object, flung from a solar system many years in the past, it was aliens sending a spacecraft through our solar system.

It was not.

The thing with a claim that anything – not just a comet shard – might be aliens, is that in order to be taken seriously, you have to do a pretty bang-up job convincing everyone that you genuinely have excluded all possible other options. And in this case, we *haven't* excluded all possible other options – we've got several pretty plausible ones still in the mix. The earliest one proposed that it was a fragment of a comet, ejected from a distant solar system, and it had wandered the galaxy until it was unlucky enough to swing close to our Sun, and then we were lucky enough to spot it on its way out.

It might also have been formed as a chunk of material that was flung out of another solar system while it was just forming planets. It could be part of something that was almost a planet, but that got too close to its star and was destroyed, sending some shattered pieces flying. Or it could be the mostly evaporated remains of some object which was once like Pluto.

People *liked* the aliens explanation, though. And so a chunk of the planetary science community has been doing damage control ever since, putting out articles with titles like "Oumuamua is not Artificial', among many others which state things equally clearly. One concludes: 'Thus, we find no compelling evidence to favor an alien explanation for 'Oumuamua.'

There are a lot of open questions about this particular visitor to our solar system. Many of them are likely to be answered only with more information, some of which is impossible, since 'Oumuamua itself has left the building. Some other answers may come with the experience we will build as we find, observe, and learn from other interstellar objects.

We still don't know *where* it came from, but it seems likely that wherever it formed, it's been traveling for a long time. And its weird shape is probably a remnant of however it was formed and flung out of that home, or a symptom of its travel through our solar system. 'Oumuamua was certainly unusual for *our* solar system. Who knows how normal these sorts of objects are in other planetary systems – and who knows when our next chance to see something as odd as 'Oumuamua will be.

IO HAS LAKES OF LAVA

If you think *you're* stressed, you should talk to Io. Io is one of Jupiter's four largest moons, and it's the most volcanically active place in the entire solar system. Like all the moons of Jupiter, Io is named after one of the (usually unfortunate) objects of the (many) amorous attentions of Zeus.*

Io itself is a touch larger than Earth's Moon, and has more than a touch more volcanoes per square meter than the Moon. There are currently *hundreds* of active volcanoes on the surface of Io, which is a sulfurous yellow-orange in color.

Moons are not normally known for tremendous volcanic activity – usually they're round chunks of rock and/or ice, either captured as they wandered the early solar system (likely true of many of Jupiter's 79+ moons[†]), or formed as a chunk of material blasted off their parent planet (as the Moon was from the Earth). Once the solar system had time to settle down into roughly the configuration we see it in now, we rather expect that moons are going to cool down, even if, like our Moon, they were formed in a violent, extremely molten smash-up.

* In mythology, Io gets turned into a cow, either by Hera or by Zeus, as punishment or concealment, depending on the version of the story. Io then wanders for a while being continuously bitten by a fly sent by an irritated Hera, and eventually gets de-bovined.

† This number keeps going up, because we keep finding more moons. Fortunately for the naming scheme, Zeus was ... prolific.

So if you're going to have a moon flinging lava more than 100 kilometers up away from its surface into space, there's definitely some other mechanism heating up the insides of that moon – our Moon doesn't do that.*

Gravity, once again, is responsible for this cosmic chaos. Io has a slightly elliptical orbit, and because Jupiter has so much mass, getting even a little bit closer to it will start to dramatically increase the gravitational force on the side of Io which faces the planet. This side is always the same: like the Moon to the Earth, Io completes one rotation every time it completes one orbit. Unlike the Moon, which takes 28 days to go around, and to rotate once, Io's in a bit of a rush, and does this once every 1.8 days.

And in doing so, it is stretched an extra 100 meters in the direction of Jupiter. The Earth's oceans do this in response to the Moon, but not by as much as 100 meters.

100 meters may not seem like a lot, but it bears repeating that this is *rock* being stretched like elastic, 100 meters in and out every 1.8 days. As we've seen before, rock doesn't *like* stretching. Its whole deal is being rock-like, not bendable at all. Stretching rock is like trying to stretch an egg, but without the option of accidentally smashing it.

So if you insist on gravitationally stretching it, a lot of the strain on the rock gets turned into heat. This process is known as 'tidal heating'. For other, more distant moons of Jupiter, the tidal heating is less intense, but it might be part

. . . .

* Alas.

of how the ice moons Ganymede and Europa can keep an underground watery ocean from freezing solid.

On Io, by contrast to its watery companions, tidal heating has made it the most bone-dry, waterless place in the solar system. There's so much heat that any water that might have been on Io is long lost – turned to gas and dissipated away. And the internal heat that is generated in this little moon comes out in the form of hundreds and hundreds of volcanoes.

Some of these volcanoes seem to be consistently active (on human timescales, anyway), some are new since we started observing, and some intermittently active. The largest of these, and indeed the most active volcano in the entire solar system, named Loki* Patera, has been an active lava lake since 1979, and really, that's just the first time we sent a spacecraft past Io,† so in all likelihood it's been around for longer than that, since 200 km lava lakes don't form overnight.

Loki Patera has some fun extra features. It seems to have a thin crust of cooler material that covers the lava lake, but every 500-ish days, something happens and the lake resurfaces itself. It may be that this is just the crust sinking into the lake, and hot lava from beneath sweeping across it. This is visible from Earth as a regular brightening and dimming of the lake – hotter, fresher lava glows brighter than the cooled-down older crusty surface. But it seems to be contained in some kind of pit.

---

* Yes, *that* Loki, the trickster god from Norse mythology. We're mixing our mythologies here. Loki has a child that's a horse (Sleipnir), so maybe they feel some kinship with Io over the whole cow thing.

† Good job, Voyager 1!

Io also has a grand total of *zero* craters. Which means that the volcanoes there are doing a spectacular job of covering up the craters extremely quickly, or we'd have seen at least a few tiny ones hanging around. Even though Loki Patera is a profuse source of hot lava on the surface of Io, it's unlikely to be doing a large amount of resurfacing, because the lava is held in the world's largest and most terrible hot tub, and not spilling out over the edges. (This is one of the reasons it's thought to be a lava lake, rather than just a pile of lava that keeps spilling out from a vent – erupting the way it has been since 1979, it ought to have largely filled in the hole it lives in.)

Fortunately, there are a few other kinds of volcanoes on Io, and these are the sort that drive 100+ km plumes above the surface of this moon. In a moment of extreme 'This isn't Earth anymore', we can divide these plume-driving volcanoes into 'giant plumes', which are the ones that stretch more than 200 km into the air, and 'generally smaller plumes', which are, you know, less than 200 km.

Just for comparison: on Earth, one definition of the start of 'outer space' is at the Kármán line, which is 100 km up.

Both of the volcanoes Pele and Tvashtar have been seen by passing spacecraft[*] to have fountained material nearly 400 km into the air – Pele some 385 km up, and Tvashtar about the same – but with enough uncertainty that it might have been anywhere between 355 and 415 km high. The tallest plume I was able to find rose faintly, but all the way to 500 km above Io's surface.

....

[*] Galileo and New Horizons, more specifically.

For some scale, the International Space Station (ISS) orbits at about 400 km above the surface of the Earth. If it were orbiting above Io at that height, it would be *hit* by a volcanic eruption. These giant eruptions are responsible for dropping new material over a 700 km radius of the surrounding region. If this were Earth, and it were centered on London, we'd cover all of Ireland, England, Wales, Belgium, the Netherlands, Switzerland, almost all of Scotland, and reach down so far south into France you'd reach Lyon and Bordeaux. If you put it in Chicago instead, you'd reach Toronto and Minneapolis, covering eight states in their entirety and all the Great Lakes except Lake Ontario. And of course it would go so far *up* that it would hit the ISS. Recall too that Io is much smaller than planet Earth, so proportionally, this is even more comprehensive coverage for Io to achieve.

These giant plumes are effectively tremendous pyroclastic explosions; the most famous of this style on Earth was that of Pompeii, where the city was buried in nine feet of ash as the result of being in the way of a superheated cloud of rock, toxic gases, and ash, sometimes described as a 'glowing avalanche', which sounds terrible (and is). So it's likely to be these volcanoes with plumes that are erasing Io's craters, since the lava lakes are doing their lakey business inside their lakey boundaries, and not overrunning their banks. It's probably the smaller plumes that are doing most of the work to erase Io's craters, since the giant plumes seem to rain down colorful circles onto Io, but they're not rearranging the surface in the same way that giant lava flows will.

We'll never send humans to Io. But we can send some

robots to orbit, and the Juno spacecraft may yet give us more information about Io in the meantime. Io is an interesting test-bed for volcanic extremes, though. While some researchers suggest that the lava on Io might be similar to that on Earth, the temperatures there are so high that it's hard to say if we should even expect it to look anything like the volcanoes on Earth. We won't know for sure until we send a robot to go check up close!

IT RAINS DIAMONDS
ON NEPTUNE

You wouldn't think that Neptune and polystyrene have much in common, but here we are. If you *very* carefully explode polystyrene, you can learn things about Neptune's interior.

Neptune is a bizarre planet. It's fantastically distant from the Earth, and has only ever been visited fleetingly, once. All the photos we have of Neptune are from Voyager 2's flyby in 1989. It takes the planet 165 years to orbit the Sun once, and light takes four hours and nine minutes to reach Neptune from the Sun; and at that distance it's too far and faint to be visible to the unaided human eye.

Neptune also has the fastest winds in the solar system, clocking in at some 1,100+ miles an hour, substantially faster than the speed of sound. We know that these winds exist, because we've watched brilliant white clouds fling themselves dramatically around the planet at a tremendous clip (both Voyager 2 and the Hubble Space Telescope have helped out with these observations).

And it's weirdly warm, as a planet. If nothing else is going on, we'd expect some kind of energy balance, with the extremely unspecific equation of:

'amount of heat a planet gets from the Sun' –
'fraction of that which is immediately reflected

back out into the cold dark' = 'the temperature
of the planet'

But Neptune is much warmer than we'd expect from this equation, which means: we're missing something.

It could be that we've got a bad estimate of how much light is reflected away, but we've sent a robot there, and it measured how reflective the planet was,* all up close and personal, so that's not likely to be the issue. The other option, of course, is that the planet itself has some way of heating itself up. It's not alone in having some extra heat source happening – Jupiter and Saturn also seem to be warmer than can be explained just by passively absorbing sunlight.

One of the ways Neptune heats itself up might be a bit unique, though. It might be dropping diamond rain down through its own atmosphere, and the friction of those diamonds ploughing through the dense atmosphere could be heating it up. Neptune may effectively be sandpapering itself warm.†

Diamond – the gemstone – is a polished and fancied-up chunk of carbon, formed when you get a lot of carbon and then mash it together at high temperatures and pressures. Theoretically speaking, Neptune has all these things. It's got high temperatures internally, because of tremendous pressure,

----

* 29% of all light hitting Neptune gets bounced back out into space.

† Very fine grit sandpaper often uses diamond as the abrasive. It's used for polishing glass, among other things. Sandpapering yourself warm is not recommended by Health and Safety.

and it's *got* carbon, though it's usually locked up in other molecules.

Neptune is mostly made up of lots of hydrogen and helium, but its distinctive blue color comes from methane, a common (and flammable) molecule among a class of what are called hydrocarbons. Hydrocarbons are simply any molecules made up of a lot of hydrogen and carbon, and nothing else. Methane in particular is one carbon atom and four hydrogen atoms, but hydrocarbons generally can have any number of carbons and hydrogens (though usually a lot more hydrogen than carbon). This is where Neptune stores its carbon. It's paired up with hydrogen, but there's a *lot* of it. The question is really: at the temperatures and pressures present in the interior of Neptune, would the carbon and hydrogen separate, freeing up the carbon to make diamonds?

We use a lot of hydrocarbons in daily life. Propane is one, which we use in gas grills and fireplaces, but the other one we know is polystyrene, or Styrofoam. We use it for packing materials, disposable plates, cheap coffee cups, etc., and it's eight carbons and eight hydrogens in a unit, and that unit just repeats endlessly* to make plastic.

Turns out, if you want to study the inner workings of Neptune without actually going there, you can fire a laser at polystyrene. Neptune is not made of polystyrene – it's hydrogen, helium, and methane, as we saw. Methane isn't polystyrene either, but it's made of the same stuff, and polystyrene has the benefit of being a solid at room temperature and also *quite*

---

* Not really. But a very long chain of this unit, in any case.

cheap. It's also less ... explosive* than methane when mixed with air at the wrong density.

It has a few drawbacks, though, and they're all 'Neptune is hot and dense, and polystyrene is neither of those things'. So: how can you make a disposable coffee cup resemble the interior of Neptune?

A Powerful Laser† enters, stage left.

In the most recent study on this, scientists shone an extremely powerful laser at some polystyrene, and then carefully studied what fell out. The laser is able to heat the Styrofoam so intensely that it abruptly turns some of it into plasma, which then effectively explodes outwards. This creates a tiny shock wave, which can create very high pressure in the surrounding Styrofoam. This kind of rude treatment turns out to be excellent at separating the hydrogen in the plastic from the carbon, and the carbon falls right out in crystal form as nanoparticles of diamond.

If you punch polystyrene hard enough, diamonds do in fact fall out.

So, by extension, if we punched methane hard enough, diamonds should fall out of that as well.

So, by extension of our extension, Neptune should be able to squeeze diamonds out of its own sky.

On Neptune, these conditions wouldn't be a tiny flash in the pan, as it were. In the lab, these ultra-high-pressure

---

* One major cause of coal mining explosions.

† A powerful laser is almost always an easy answer when the question is 'How to make this material hotter and/or denser'.

conditions are only able to be sustained for tiny blips of time. On Neptune, they'd be chronic. So it's possible that the nanometer-sized diamonds they were able to form in the lab during their brief laser-blasting could keep growing to larger sizes.

The thing with diamond rain,* of course, is – where is it going? 'Down' is not very specific. Do you have a surface to land on? Is it a liquid layer? Is the diamond collecting down in the core of Neptune? Is there a diamond layer in Neptune?† Are there diamond icebergs?

To learn more about Neptune we'll have to go back, some-day. But in the meantime, these studies of the oddities of our own solar system's planets, once again, inform how odd we might expect other planets in other solar systems to be. Assessment so far: very odd indeed.

....

* One of many things.

† Don't get your hopes up that we can retrieve any of it. Trying to get any-where near these crushing pressures and temperatures is not a great idea for any mechanical craft, and even less of a good idea for a human. Much easier is to make your own diamonds here on Earth, which we can already do.

AN EXOPLANET WE THOUGHT WAS MADE OF DIAMOND MIGHT BE LAVA INSTEAD

There's a class of exoplanets (i.e. planets outside our solar system) that formed around stars which are particularly full of carbon. And if the stars are full of carbon, then it's reasonable to assume that the planets, forming out of the same gas cloud, should be full of carbon as well. If you happen to have a massive enough planet, the theory goes, you should be able to take all that carbon and crush it into a giant layer, much like the diamonds in Neptune's atmosphere. And indeed, if you try to replicate high-pressure environments in the lab, we have confirmed that if you smash silicon carbide in water between two diamonds, you get more diamond and silica, so this process could well happen inside an exoplanet.

We humans do love ourselves some diamonds. While it's not an exclusive fascination (we generally enjoy all sorts of sparkly rocks), diamonds have a particular cultural place for many humans. When we find the presence of diamonds elsewhere in our Universe, the sparkly rock portion of (at least) my brain is particularly pleased by the delightful impracticality of a planet that might be one-third diamond.

But the exoplanet which was most excitedly proclaimed to be Diamond Planet (55 Cancri e) may not be, after all. Don't worry: we've simply replaced a giant diamond with a *giant ocean of lava*.

The journey in our understanding from 'diamond planet' to 'ball o' lava' is a wild one, and it's a great illustration of how hard it is to learn about the surface and atmosphere of a planet 41 light years away from Earth.

55 Cancri e is the fourth planet discovered around the star 55 Cancri, and it's profoundly roasted by its parent star, having a surface temperature of 2,700 Kelvin. The star itself isn't that different from the Sun – they're nearly identical – but 55 Cancri e orbits it once every eighteen hours, which means it's *very* close to the star. For comparison, Mercury's orbit around the Sun takes 88 days, and it's nearly completely bone-dry, with a typical surface temperature of 430 C.

55 Cancri e is twice as physically large as the Earth, and eight times more massive, which tells us it's substantially more dense than our home. In fact, overall, it's about as dense as lead. Now, the planet is unlikely to be entirely made up of lead, as lead is relatively uncommon, so in trying to figure out what it's *actually* made of, we have to mix some denser and some lighter materials, and average our way out to 'very dense'.

On top of all that, it'll be tidally locked – one side of this planet is permanently facing the star, and the other side permanently facing away from it. And, as we observe it, the day-facing side of the planet is substantially warmer than the night side, unsurprisingly.

Thus far, everyone is agreed.

After this, it gets a bit tangled.

There are a few broad classes of planet composition that

you can have for a planet that is eight times more massive than the Earth.

One: it's a ball of rock, with no atmosphere, like an enormous Mercury.

Two: It's a ball of rock, but with a thick atmosphere around it – more like Venus.

Three: It's a ball of rock surrounded by dense, hot water, compressed into liquid form by its intense atmosphere, unlike anything in our solar system, but more like a warm Neptune than anything else.

Four: It's a ball of material that normally we wouldn't call *rock*, but is dense and solid, and might or might not have an atmosphere, also unlike anything in our solar system.

It's this fourth category that brought us the temporary delights of a diamond planet. Diamond *is* very dense, and if the planet were made of enough carbon, it's possible.

There's a lot of variability here, but ultimately folks have been struggling with three questions: 'Does it have an atmosphere?'; 'How wet is it?'; and 'Are these normal or weird rocks?' – and to be honest, *most* of exoplanetary science folds into these three questions.

In the case of 55 Cancri e, it seems as though the answers are settling towards: 'Yes, atmosphere'; 'Not wet'; and 'Probably medium normal rock?' – which leads us to envision a world more like a giant and extreme child of Venus and Io.

In general, the more extreme the temperature difference between daytime and nighttime, the more unlikely a planet is to have a thick atmosphere; atmospheres hold on to heat, and temperature gradients can give you a wind, so if you measure a planet and it's the same temperature on both sides, even though only one side can be facing the star at any given time, then you have a really efficient system of spreading out the heat it receives. Rock is *not* an efficient system, even when molten, so usually the fingers go pointing to 'thick, circulating atmosphere', which is doing the work of a central heating system, pushing warm air into cooler spaces.

In 55 Cancri e, there's a strong temperature difference – the star-facing side is usually about 1,000 degrees warmer than the night-facing side – but counterintuitively, this isn't as strong a signature as you'd expect if the planet were bare rock, and there are some signs that the planet is hottest *not* right where its star is shining directly above it, which means that something is shifting the heat to one side – probably an atmosphere, but a slow, sluggish atmosphere, made of heavy molecules like carbon dioxide, carbon monoxide or nitrogen. An efficient, easily moved atmosphere would mean that the day/night temperature switch would be less intense.

But just because molten rock can't account for the temperature swings from day to night doesn't mean it couldn't still be there, underneath the thick atmosphere. And it very well might be: although 2,700 Kelvin is hot enough that molten lava would behave in a way 'more similar to water at room temperature than to solid rock', the night side, on the other hand, is cool enough that the surface could be solid. An open lava ocean also provides a way to create such a dense atmosphere. The atmosphere that is obscuring our view of the planet's surface could be a combination of volcanic gases bubbling up from the depths, and some of the lava itself evaporating out into the atmosphere. Any surface lava, dealing with both the star's heat and the scorching temperatures of the atmosphere itself, could well be heated into a gas.

So, this is the entrance of our 'planet which is half lava ocean' idea. If the atmosphere is also part evaporated lava, and it will cool as it moves to the night sky and the relative shelter of the night side of the planet, you could reasonably* ask: 'Does it rain lava?' Future telescopes may be able to tell us. If it does, then the atmosphere will be full of silicate material evaporated out of the lava oceans – and if this begins to fall out of the sky as rain, then we have this delicate vista, suggested by NASA, to imagine, hidden away from the intense heat and glare of the star and the liquified rock ocean: 'Silicates in the atmosphere would condense into clouds on the tidally-locked planet's dark side reflecting the lava below. So, the skies would sparkle.'

In the case of 55 Cancri e, we may have exchanged a crystalline interior of a diamond planet for one with crystalline skies.

---

* Well, 'reasonably' may feel like a stretch, but it's following a logical path, anyway.

THERE'S A PITCH-BLACK EXOPLANET

Zooming around its host star once every 26 hours, in the constellation of Auriga, a planet as black as tar is losing a slow gravitational fight.

The planet itself bears the name of WASP-12b; WASP after the survey in which it was found (the Wide Angle Search for Planets*), and 12 designates the host star. Lower-case letters indicate the planets, and so 12b is the first detected planet around this system. 12B would be another star; and WASP-12 (the primary star) does seem to have two red dwarf companions, thereby named 12B and 12C. (Have we mentioned that all astronomical naming schemes are temporary hot fixes piled on top of each other, and accidentally made permanent?)

At the time of its discovery in 2009, WASP-12b was the hottest known exoplanet. It was also the largest, and had the shortest orbital time.

The host star itself is very like our own Sun, just a bit more massive† and a little hotter,‡ though much younger – it's only around 2 billion years of age, compared to our Sun's 4.5 billion.

. . . .

* Not as bad as it gets, to be frank.

† 1.3 solar masses.

‡ The surface temperature of the Sun is 5,780 Kelvin, and WASP-12 is 6,300 K.

It's also brighter: small changes in mass and temperature correspond to big changes in the amount of light it churns out, so even though it's only 30% more massive than the Sun, this star is nearly 3.5 times as luminous.

The planet, on the other hand, is unlike anything in our solar system. It's 1.8 times the physical size of Jupiter, but only about a quarter as dense, making it a huge, puffy planet. And it's *very* close to its star; while Mercury orbits our own Sun at 0.4 au,* WASP-12b orbits more than ten times closer in at 0.03 au. It's being *broiled* by its star. Its estimated temperature sits around 2,500 Kelvin, hotter than some failed stars.

How can we tell how dark a planet is? After all, we can't go there directly, and when something absorbs 94% of the light that hits it, as WASP-12b does, it's not exactly going to be a bright object in the sky.

It's a clever trick of how we detected it in the first place.

If you're very lucky, and the geometry gods have smiled on you, and you also have a technologically advanced society able to build incredibly sensitive cameras, you can point those cameras towards a large number of stars and wait to see if the amount of light from them flickers. There are a lot of reasons why the light from a star could flicker, not least that many stars are unstable and produce an irregular amount of light. So if you're looking for a planet, you want that flicker to repeat on a regular, predictable cadence. This will only occur if the planet (from our perspective) passes in front of the star, which is why you need the smiles of

---

* au = astronomical unit, defined as the distance from the Earth to the Sun (roughly 150 million km).

the geometry gods. There are a lot of ways to orient an exoplanet around a distant star where this simply doesn't happen.

So if you detect a repeated drop in the amount of light you receive from a distant star, you can be reasonably certain that *something* is blocking light from that star. The problem is that the culprit could also be a dimmer star, as a lot of stars are found in double systems. To rule out another star, you'd either need the mass of the orbiting object, or to get an estimate of how much the light is dropping by.

So let's say we did that – we have a star, with a planet passing in front of it, and we can detect the dip in the amount of light received. The dip is larger the physically larger the planet is; larger planets simply block more light. And the faster the planet orbits, the more chances you have to see it passing in front of the star. Once every day is a lot easier than something that only crosses its star once every four months.

When the planet is between you and the star, the starlit side faces away from you, and we would not expect much light in the visible wavelengths to come from the night side of a planet. (This is only true for visible light; if you're looking in the infrared, which measures heat, you might still see a glow.) But, when the planet is about to transit behind the star, we should be seeing both the star itself and the reflection of the star's light off the surface of the planet back at us.

Scientists attempted exactly this measurement with the Hubble Space Telescope. Since we already knew the size, mass, and orbit of WASP-12b from previous observations, they were just looking to see how much light the planet would reflect back towards us.

The answer: almost none.

To reflect only 6% of the light that strikes it, the remaining 94% must be absorbed. This absorption was independent of color — and the only way you get that is if your planet is black.

It could have been the case that this planet absorbed lots of red light and reflected blue, in which case we would have concluded that it was a very blue planet. If it absorbed everything except the darkest red colors, we would conclude that this was a very dark red planet. But when you absorb *all* optical light, that there is a pitch-black planet.

The blackest artificial substance to date is a vertical forest of carbon nanotubes, and it absorbs 99.995% of all incoming light; so our planetary find isn't the *least* reflective thing we can imagine. But with an upper limit of 'no more than 6% light reflected', the planet is as dark as freshly laid asphalt, or the open ocean. Its opposite — a substance that reflects 90% of the light that strikes it — is fresh snow.

AN ACCURATE RENDITION OF WASP-12b

How do you make a planet so dark? That's still a mystery. We've observed this planet in nearly every possible wavelength, and it's effectively a pure black body – meaning that it absorbs absolutely all light that strikes it. The other extreme form of black body is a star, which absorbs all incoming light because it's glowing intensely: you'll never get a reflection off a star. Where a star is bright, this planet is dark, glowing only in the infrared due to its own heat. One suggestion is that its atmosphere may be full of dust; tiny, micron-sized flecks of dust in the atmosphere would absorb light. It is one of the universal truths: dust gets in the way of seeing what's behind it.

Besides the fascination of trying to work out how you make a planet so dark, planetary scientists are trying to figure out how you can get such a large planet to orbit its star so closely. In fact, it's not doing a great job of orbiting its star safely – it's being shredded. The planet is notably elongated, stretched towards the star it orbits, and so large that its thin, outer atmosphere is more strongly pulled to its star than it is to itself. On top of that, the orbit of the planet is thought to be decaying – the planet is falling into the star wholesale, and is expected to last only another 3 million years. Since the parent star itself is only around a billion years old, by the time this planet falls into its atmosphere, or is shredded into gas and then falls into the atmosphere, the planet will not have lasted as long as the Earth has already existed, and we are catching this planet in the last gasp of its tar-black, gravitationally distorted existence.

THE MOON SMELLS
OF GUNPOWDER

After a moonwalk in April 1972, the Apollo 16 astronauts Charles Duke and John Young returned to their capsule. In the process of putting their suits and other things away, Duke commented to Ground Control:

> 150:09:18 Duke: Houston, the lunar dust smells like gunpowder. [Pause]
>
> 150:09:27 England: We copy that, Charlie.
>
> 150:09:31 Duke: Really, really a strong odor to it.

Now, obviously these astronauts weren't taking their helmets off and taking a big whiff of the non-existent lunar atmosphere; that's a way to a nasty death. They were safely back in the lunar module and just sniffing the air. For all their care in collecting specimens, the lunar dust had clung to their suits; and, shaken loose once returned to the craft, it was free to float around the module and get in everyone's noses. And into their mouths:

> 126:10:40 Young: Hey, that Moon dust don't taste half bad.
>
> 126:10:44 Duke: Is that what that is?
>
> 126:10:45 Young: Yeah.

Apollo 16 wasn't the only mission that noticed this – 'gunpowder' was a frequent description, and every single mission that landed had issues with dust floating around the spacecraft after returning from moonwalks.

Apollo 17, the last of the Apollo missions, was the only one to carry a geologist to the Moon in Harrison 'Jack' Schmitt, who had a similar exchange with his crewmate Eugene Cernan. Returning from a moonwalk, he commented on the dust:

> 124:19:47 Schmitt: Smells like gunpowder, just like the boys said.
>
> 124:19:53 Cernan: Oh, it does, doesn't it?

Schmitt wound up being a little allergic to it. Ground Control teased him about sounding congested:

> 127:47:21 Allen: Sounds like you've got hay fever sensors, as far as that dust goes.
>
> 127:47:35 Schmitt: It's come on pretty fast just since I came back. I think as soon as the cabin filters most of this out that is in the air, I'll be all right. But I didn't know I had lunar dust hay fever.
>
> 127:47:54 Allen: It's funny they don't check for that. Maybe that's the trouble with the cheap noses, Jack.
>
> 127:48:05 Schmitt: Could be. I don't know why we couldn't have gone and smelled some dust in the LRL [Lunar Receiving Lab] just to find out.

This is one of the most unusual facts we have about the Moon, and we wouldn't know it except for having sent humans there. We know the chemical composition of the Moon's dusty layers from observing it from orbit, but knowing that it additionally smells like gunpowder hasn't really advanced our understanding of the Moon by very much.

However, the fact that it was so easily tracked into the lunar module and inhaled *did* change our understanding of how humans will have to operate if we are ever to live on the Moon for longer than a couple of days at a time. The lunar dust is not great for the respiratory system.

Dirt on Earth is usually not very sharp; small pieces of rock and degraded plant material are tumbled against each other and generally turn out somewhat polished, like river rocks, before they enter our noses. If you happen to be allergic to dust, it's bad luck, but it's not doing much in the way of physical harm.

The lunar dust, on the other hand, is the shattered remains of rocks, broken repeatedly by tiny meteorites striking the surface. It's *sharp*.

So sharp, in fact, that it slashed the seals on some of the vacuum-sealed bags meant to preserve Moon dust on the way home; they wound up being contaminated with oxygen by the time the Apollo missions made their three-day trip back to Earth.

It clung so severely to the moonwalking space suits that even brushing each other off before returning to the module effectively did nothing to remove the dust. Considering that the astronauts were notoriously clumsy on the lunar surface, trying to adapt to both the unwieldy suits and the lowered gravity, most of them had taken several tumbles over the course of their moonwalks, and these suits were no longer pristine after many hours on the surface. They were, instead, rather comprehensively covered in lunar dirt.

It was more than just getting wedged in the folds of the suit – it was static cling. If you have ever seen a cat try to extract itself from a box of packing styrofoam without trailing pieces stuck to all parts of itself, that's the problem we were having with the lunar dust on the Moon.

Even without falling down, they were kneeling to collect samples, so the dust was pervasive, and generally problematic. But if it were just an irritation, we could find ways to keep the dust from clinging so severely,* and quickly destroying air filters with tracked-in Moon dust.

There's a more pressing need for this technology, and it's to protect our squishy human bodies.

The human lung does not *like* tiny microscopic shards of rock. Breathing these in can damage lung tissue in a way that is difficult to repair, because the rocks are so sharp and so tiny that simply coughing won't expel them, and so they stay embedded in the lungs, continuously doing damage and eventually causing problems similar to very severe pneumonia.

There's an earthbound parallel called silicosis, which comes from breathing fine mineral dust, most notably from mining quartz, and which still causes deaths today, less now from mining and more from the cutting of quartz countertops without

* This is a thing we're doing anyway! A thin coating of indium tin oxide may be used as a paint or a general coating to help shed the static cling.

proper protection. Between 1999 and 2019, 2,512 people in the US died of silicosis. Like the Moon dust, quartz isn't intrinsically *toxic*, it's just that it's like inhaling fine shards of glass, which isn't a great idea.

But it's one of many problems we're going to have to solve if we want humans to go live on the Moon. Sending them off to a land of Stark Beauty And Irreversible Lung Impairment is not quite the same proposition as sending them off to a land of Stark Beauty And Vistas Of Earth.

It's also critical to learn how to do this if we ever want to send humans to Mars. The dust on Mars has the same 'like breathing glass' problem, but in addition to that, it's actively bad for the human body. While the Phoenix lander showed that the soil was not incredibly acidic or basic,* the finely-ground rocks present there *do* react strongly with water. They create hydrogen peroxide, at much higher levels than quartz dust, which is one of the problems you run into with silicosis. The authors of one study remark that the dust on the Moon and Mars could 'potentially creat[e] more deleterious effects', which is technical speak for 'it's almost definitely tremendously bad for you'.

This is not something you'd want in your lungs, on your skin, or generally anywhere near you. But if we can sort out what to do with the Moon dust, maybe by the time humans are ready to go to Mars, they'll be more prepared to deal with Mars' equally fine, shard-like, clinging dust.

---

* Bases are, in part, defined as dramatically reactive if you mix them with an acid. The best-known of these pairings is baking soda (bicarbonate of soda), which is a base, and vinegar, which is a mild acid.

YOU COULD GROW TURNIPS ON MARS SOIL IF IT WEREN'T FULL OF ROCKET FUEL

When the Phoenix Mars lander arrived at the surface of Mars in 2008, it brought with it an elaborate chemical laboratory ready to ask deep questions like: 'Is the surface of Mars acidic or basic?'

This was an important first step to ask, because humans don't like exposure to either acids or bases, though we often have both in our houses. Bleach, for instance, is a base, and if it touches your skin, you're very likely to get a chemical burn. On the other end of the spectrum, lemon juice is an acid (citric acid, in isolation from the fruit) but a lot of acids are more dangerous – such as sulfuric acid.

And it seemed so promising at first! Phoenix returned soil samples that said that the soil was not acidic! It was just a little bit basic.

Okay, maybe more than a little basic. These things are measured on the pH scale from 0 to 14, with water in the middle as a neutral 7. Anything less than 7 is an acid, and anything greater than 7 is a base. Lemon juice is a 2. Bleach is an 11.

The Martian soil where Phoenix landed was between 8 and 9. This is somewhere between 'seawater' and 'baking soda' on the pH scale.

It also contained a lot of nutrients that we know plants need to grow, and while not all plants can grow in slightly basic soil,* some can, and so we got lots of headlines like 'Martian Soil Could Grow Asparagus'† or 'Martian soil could grow turnips, Phoenix finds'. Most fruits need acidic soil, so strawberries were out, but simulated Mars soil got leafy greens (lettuce and kale) to grow in greenhouses on Earth.

Unfortunately, the Phoenix lander also found another compound shortly afterwards: perchlorate. One form of perchlorate (ammonium perchlorate) is used in rocket fuel – this is what was in the solid rocket boosters that helped launch the space shuttles into orbit. In general, it is my life's mission to eat as little solid rocket fuel as possible.

It turns out that medical science agrees with me. All perchlorate compounds (regardless if they're ammonium perchlorate or some other form) are *tremendously* bad for you! They interfere with the proper function of the thyroid, which is not one of the optional parts of the human body.‡ If you have too much of it in your system, it is toxic, and it can make you dead.

Mars has it everywhere.

And not in a small amount. Phoenix reported that the soil where it landed was 0.5% of perchlorate salts, likely calcium

---

* The potato, for instance, likes acidic soil. Sorry, Mark Watney of *The Martian*.

† Though Jim Green of NASA is quoted as saying: 'If I had to eat asparagus for three years, I think I'd just take my helmet off and walk outside.'

‡ Looking at you, the appendix.

and magnesium perchlorate, and that is *way too much perchlorate* if you are to remain a healthy human being.

The Curiosity rover found perchlorate in dirt it analyzed, and it has also been found *from orbit*, by tracing the chlorine it bears. Perchlorate is stable in water, so even the water that we can see from orbit is likely tainted with toxins to humans.

And it's not just humans that perchlorate is bad for; it will kill a lot of things stone dead. If you take these concentrations of perchlorates, and expose them to the amount of UV light that Mars receives, you can kill off bacteria in minutes. If you additionally include the fact that Mars dirt has peroxide (color-safe bleach) and iron oxide (rust) in it, and run the experiment again, you can kill bacteria *ten times faster*. RIP *Bacillus subtilis*, we hardly knew ye.* Researchers conclude that 'the surface of present-day Mars is highly deleterious to cells, caused by a toxic cocktail of oxidants, iron oxides, perchlorates and UV irradiation'.

'Highly deleterious' is what scientists put in technical papers because editors of technical journals won't let you write: 'Mars is great as long as you don't want anything alive to stay that way.'† Mars as a whole has perchlorate levels between 0.5 and 1% of the soil, pretty much anywhere we look, which is 1,000 to 10,000 times higher than what we get on Earth.

....

* This particular bacterium has a penchant for surviving clean rooms, and is therefore a persistent jerk of a bacterium. To be immediately killed by this experiment tells you something about exactly how bad for you this chemical environment is.

† It's too bad. It'd really liven up a paper.

Just touching the soil wouldn't poison you, but breathing in dust, which would already be bad for you, would be able to deliver this chemical directly into your body. A few *milligrams* of dust would be more than the recommended dose of perchlorate.

Also concerning: if you let UV radiation pummel your perchlorate, as it would on Mars, it can degrade into *even nastier* compounds (ClO_2^{-*} and $ClO^{-\dagger}$) which are straight-up corrosive, and come with the possibility of fun 'health concerns such as respiratory difficulties, headaches, skin burns, loss of consciousness and vomiting'. Both of these are used as disinfectants on Earth, but disinfecting the inside of the human body has a habit of making you free of living cells, aka *very dead*.

Unlike bacteria and the human thyroid, some plants seem to be able to grow with Earth-levels of perchlorate around. The problem is that plants, like the human thyroid, absorb perchlorate, so if you're growing plants with the intent of eating them, congratulations, you've made poison plants out of poison dirt. For extra fun, the perchlorate tends to accumulate particularly strongly in the leaves, otherwise known as 'the bit you normally want to eat'. So any Mars Spinach you'd be able to convince to grow in the soil there would be Poison Salad for an astronaut if you haven't dealt with the perchlorate first.

Of course, these experiments are with Earth-like levels of perchlorate contamination – the levels you can find around factories that manufacture rocket fuel and fireworks – and this

----

* Bleach. It's just bleach. Do not inhale it. It will murder you.

† Sodium hypochlorite, aka pool chlorine, is very bad for you.

is a lot lower than what you'd find on Mars. So what happens to plants when you ramp up the perchlorate?

They die.

Simulated Mars dirt can vaguely manage to support plant life if you fertilize it enough, but the more lifelike you make your soil, the harder it is to grow things in it. In this particular model, researchers began with three different types of fake Mars dirt: one from a cinder cone in Hawaii, one from the Mojave desert, and the third created entirely from scratch to try and match the properties of Mars soil as closely as possible.

They planted romaine lettuce and mouse-ear cress in the three Mars dirts, and the one built from scratch just murdered the seeds, even without the perchlorate, and even with fertilizer. They tried getting the seeds to sprout hydroponically (i.e. in water), and then transferred them into the custom-built Mars dirt, and they also promptly* expired.

In this case, it seems that one of the major factors was that the soil was too basic for the plants to survive. This soil had a pH of 9.5, which is a bit higher than what was measured by the Phoenix lander, so to fix that, they poured sulfuric acid over the dirt,† which altered the pH back towards something more neutral. They got plants to survive for a grand old extra week, and then they all died.

The desert and volcano Mars dirts did manage to get the plants to sprout, but without adding fertilizer, both plants died

---

* In five to eight days.

† Yeah. This sounds like a bad idea, and for sure you should do this *before* you plant your seedlings.

within a week. With fertilizer, but without the perchlorate, they summarized their results thusly:

Simulant	Edible biomass
Volcano dirt	21 g +/– 4 g
Desert dirt	18 g +/– 5 g
Custom-built dirt	DEAD

Add in the perchlorate, though, and everything dies. Their mouse-ear cress gave up the ghost in less than five days, and the lettuce was gone in nine days. This table wasn't in the researchers' paper, but I can summarize their results of trying to grow plants in perchlorated soil, after a week of being in there:

Simulant	Growth stage
Volcano dirt	DEAD
Desert dirt	DEAD
Custom-built dirt	DEAD

Mars dirt, it seems, is not great for plants, bacteria, *or* the human thyroid.

Just what you want on a long mission from home: a malfunctioning thyroid, dust that can shred your lungs *and* corrode through materials, and a planet with a blacker thumb than any human who has bad luck with house plants.

THE MOON ONCE HAD
LAVA LAKES
AND FIRE FOUNTAINS

It's 3.5 billion years ago; the Moon has only been around for about 1 billion years. It's closer to the Earth by about one-third (135,000 km closer), and the first creatures on Earth whose fossils have been uncovered are growing in our oceans.

The Moon is about to be on fire for the next 2 billion years or so.

It's already made it through a rather difficult period called the Late Heavy Bombardment,* which ran from about 4 billion years to 3.5 billion years ago, a time that produced many Very Large Craters on its surface. The internal workings of the Moon are still much warmer than today, which means its interior is still more molten than it is now. In turn, this means there's magma on the Moon.

There's been a lot of evidence for extinct lunar volcanoes over the years, and this evidence was aided tremendously by the retrieval and study of Moon rocks by the Apollo astronauts. But we've only been there six times, and six landings is not enough

---

* Late, because it was after most of the planets had formed, and Heavy, because the craters are large, so the objects which struck it were Also Large. Bombardment for 'things hitting other things'.

to get all the interesting rocks,* so a lot of our understanding of the Moon's volcanic history comes from imaging it from orbit, rather than landing there. So: how can we set the Moon ablaze?

Much like volcanoes on Earth, the story began with magma rising to the surface. However, very much unlike volcanoes on Earth, the Moon's volcanoes were not affiliated with plate tectonics. Instead, blade-like upwellings of lava rose upwards from deep in the Moon, traveling along old fractures in the lunar rocks, or creating their own if they could. Many of these blades of magma never made it to the surface; they didn't exert enough force to reach it. For magma rising towards the far side of the Moon, this was nearly always its fate – the far side has a thicker crust, and many of these would-be eruptions stalled out before they could break through.

On the near side, though, they were able to break through frequently enough that, eventually, lava covered 17% of the entire lunar surface, or around one-third of the near side. All of the dark patches on the Moon, visible by eye and known as *mare*,† are the remnants of lava flows, dating somewhere between 3.5 billion years and 1 billion years ago. Many of these mare were probably impact craters from the Late Heavy Bombardment before they became ideal low-lying areas to fill in with lava.

. . . .

* I'd be rich if I had a penny for every time I saw a very technically phrased version of 'IF ONLY WE COULD GO PICK SOME ROCKS OUT FROM HERE SPECIFICALLY', in science papers about lunar geology.

† This is the Latin for 'sea' – the plural of mare is *maria*.

Volcanoes on the Moon were a touch dramatic, and they left a lot of clues for us that they must have been spectacular — the first being tiny glass beads *all over* the surface. They're present in pretty much every chunk of dirt that the Apollo astronauts brought back, and the way you make those is to *explosively fling molten lava* out of a volcano.

In fact, the way that volcanoes erupted on the Moon had all the grace and nearly the same mechanism of a shaken soda can, except instead of abruptly dousing the surroundings with sticky sugar water, you're going to have a seething foam of lava.

An unshaken soda can doesn't foam over when it's opened, because the gas that creates the bubbles is dissolved in with the liquid, and it (normally) takes some time to create a bubble and rise to the top. But if you leave it long enough, your soda can go flat; all that gas has escaped, and the carbonation has disappeared.

Magma, in its depths, also has gas dissolved in with the molten rock (our stand-in for soda). On Earth, two things keep every eruption from explosively decompressing: one is that generally the lava rises slowly, and so bubbles of gas don't experience a dramatic change in pressure (for our soda can, this pressure change is created by opening the can). And the second is that we have a whole atmosphere of air pressure, so even when there *are* bubbles, their expansion is resisted at the surface of the Earth by the entire atmosphere of gas pressing down on them.

3.5 billion years ago, when the lunar volcanoes were active, the magma blades rose extremely rapidly. In fact, it only took 4.6 hours to rise from a depth of 500 km, which is about

100 km/hour – that's highway speeds. In addition, the Moon has no atmosphere, so when the rising magma reached the surface, it suddenly experienced a vacuum, instead of the pressure of the interior of the Moon.

Because these rising daggers of magma ascended so quickly, bubbles also formed rapidly, and they grew in size as the pressure from the Moon's crust decreased. Bubbles were less dense than the surrounding magma, so they rose within the column of magma, and collected near the top of it. By the time the magma knife reached the surface, what we actually had was a pointed tip entirely of gas, followed swiftly by a foam of mostly bubbles of magma.* Beneath that was more 'traditional' magma, mostly molten lava with gases still dissolved within it.

So when the tip of this rapidly rising hot material first reached the surface of the Moon, what encountered the vacuum of space *first* was a pocket of highly pressurized gas.

This exploded.

Very literally, this was a bomb going off, which it did as soon as the first gas was able to escape out from the underbelly of the Moon. There's a calculation for how big bubbles can expand based on the pressure differential, but when the pressure you're moving into is Zero, Because Space Is A Vacuum, the bubble should reach infinite size, which ... is impractical, but definitely means the bubble will expand rapidly, dramatically, and explosively.

This is how you get glass beads scattered all over the Moon. This explosive first stage blasts surface material away

---

* Think the worst bubble bath ever.

from the new hole in the ground, and exposes the magma foam to the dark void of space, so it explodes too. But the foam has little bits of magma (now lava) clinging to it, so when these bubbles explode, they drag little bits of molten rock with them, raining them down broadly over the surrounding 5–10 km of lunar surface. Those bits of molten rock will cool into tiny glass beads – the same sort the Apollo astronauts were finding everywhere they looked. If you repeat this explosive detonation process enough times in enough different locations, it's no mystery you could fling little glass beads all over the near side of the Moon.

That, of course, was not the end of the volcanic drama, but only the beginning. And lunar volcanoes, compared to Earth ones, could be Voluminous. (This isn't true on average, but when the Moon had a big eruption, it was more on par with the most catastrophic volcanic eruptions the Earth has ever seen. Smaller eruptions were more common, but not as dramatic.)

If there was enough lava coming out of this hole in the ground at a fast enough pace, and typically there was – at somewhere between 10,000 and a million cubic meters of lava every second – then the pieces of lava getting flung out of their foamy carbonated mess would be thick enough that the void of space couldn't cool them down effectively. The outer layers of the cloud of boiling lava would certainly cool, but the inner layers could stay so hot that when they landed back down on the ground (particularly near the vent) they were still at the temperature of Boiling Hot Lava. On Earth, if you fling lava up into the air and it lands still molten, it's given the name a 'fire fountain'.

Do this persistently, and you would build up a lava lake where these molten shards of rock were landing. Theoretical models (and the remains of these lakes on the Moon) tell us they'd have been somewhere between a few hundred meters and 2,000 meters across. None have been found more than 4,000 meters in radius.

Disappointingly, since the lava lake only exists because of a thick cloud of exploding gas keeping the erupting material very hot, it's very unlikely it would have been visible from the early Earth to any adventurous time-traveler.

If you keep raining down molten lava onto the ground, your lava lake will start to drain downhill, wherever that might be, and now you have a lava flow, fueled by the lava continuously flung out of the vent and into the lava pond. The massive craters that we now see as volcanic mare were easy targets, as they are easily 'downhill'. (They were also easier to reach by rising magma, because the divot they had formed was deep.)

Most eruptions were small overall, but the rate at which lava was pushed out onto the surface was *rapid*, so the lava flows were, relatively speaking, speedy. A 600 km-long lava flow, 30 meters deep and 20 km wide, could unfurl in a little over a day.* But if it was only 1 meter deep instead, by changing the properties of the lava, it could have taken up to 11.6 days. A 1,200 km-long lava flow† could very plausibly

---

* It's about eight times slower than driving that same distance, but more direct and much more destructive.

† Lava flows less than 1,200 km ran out of lava before they'd start to cool down.

have taken a little under three days* to wend its way across the lunar surface.

But some volcanoes, which had a particularly large volume of lava rise to the surface, could keep going for nearly four months, eventually feeding what was effectively a river of lava (technically called a sinuous rille).

The sheer number of times the Moon cracked open and leaked lava eventually laid down 10 million cubic kilometers of lava. A lot of that goes deep, rather than wide – it seems that in some places, the lava layers must be close to 7 km deep, though on average they are more likely to sit around 1.5 km in depth. Considering that, typically, an eruption only laid down tens to hundreds of meters of lava,[†] this leaves room for *a lot* of volcanoes over the lunar history. Probably 30,000–100,000 of them. Each with their own lava pond, each with their own lava flow or river, depending on how much lava needed out.

Given that we have 2 billion years to fill, the odds are good that there was a lot of time between each eruption,[‡] even knowing that they were most frequent and most prolific about 3 billion years ago. Large[§] eruptions probably occurred every

. . . .

* Three to four days seems about typical for how long it took the magma vent to empty itself nearly fully.

† One of the ways we know this is by looking at craters that have punched through one layer of lava and revealed a chemically distinct layer below, just like punching your fist into a cake will reveal how deep the frosting layer is, and whether or not the cake is chocolate.

‡ By the estimates in the paper referenced in the endnotes, 20,000–60,000 years or so.

§ Defined as 1,000 km^3 of lava laid down in a year.

million years or so, which leaves room for about a thousand of them. If it only takes a year or two for the lava to cool completely, by the time the next eruption turns up, it's flowing over cooled lava, in all likelihood.

If you're disappointed that we humans missed all the fun volcanoes on the Moon, you'll be delighted to know that while we definitely missed the impressive bit, it seems as though the Moon remained a tiny bit volcanically active for a long time afterwards – there are hints that very small eruptions continued to occur as recently as 100 million to 50 million years ago. These dates are hard to pin down, because we haven't got any rocks directly from these zones to date them specifically, and when you're trying to estimate an age from afar, the most common method is by counting the number of small craters peppered around the surface. If there aren't any craters around to be seen, then you really can only conclude that the surface there is 'fresh'.

100 million years ago, taking the older date from an abundance of precaution, puts a volcanically active Moon at the time of the dinosaurs, who surely didn't appreciate it at all, given the impromptu volcano* that triggered their own demise.

----

* If you punch a crater big enough in the Earth's crust, you make a sudden lava tsunami, which, though it would make for an excellent band name, is not the best way to have an uneventful Tuesday.

SATURN'S LESS DENSE
THAN WATER

Saturn is the second most massive planet in our solar system, after, of course, the tremendous Jupiter. But while Jupiter has about three times the mass of Saturn, it's only 1.2 times as physically large, so Saturn, by comparison, is pretty fluffy. Not very much mass, but *large*.

To scale, the two planets look like this, if you skip Saturn's rings.*

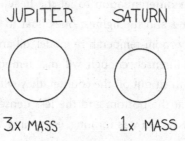

Eight times less dense than Earth, Saturn is also substantially less dense than all of the other gas giant planets. In fact, Saturn is the least dense planet in our solar system. It's a weird outlier, this fluffy ringed planet: it is the only planet in the solar system whose density, on average, is less than that of water.

....

* Saturn's rings are also *enormous*, and would make this comparison harder to see.

'On average' is doing a lot of work here, because the density of planets is not usually uniform, and that's certainly true of the gas giants, which have *very* fluffy outer atmospheres, and very high-pressure and high-density cores, where the gas they're made of is compressed into behaviors we don't normally see here on Earth. The only objects which *do* have a uniform density are usually things like asteroids, which are too small to have sorted themselves into a density onion with the most dense materials at the center, as we see with all the other larger objects in the solar system.

This process of self-sorting occurred early in the formation of the solar system, and is known as differentiation. Rocky bodies tend to need to have spent at least a little time molten hot in order for differentiation to let the heaviest elements sink to the center and the lightest ones float to the top. For a quick and easy room-temperature model, oil and vinegar, if you shake them intensely enough, will mix, temporarily. But if you leave them to sit out on the counter, they will settle, with all the vinegar at the bottom and the less dense oil on top. When the solar system was forming, everything was more or less in a shaken-and-very-mixed state; but leaving the planets as a molten ball for a while allowed them to settle out. If the rocky bodies were never warm enough to settle like this, then effectively they could be frozen the way they were at the start. Objects, like asteroids, which were either never warm enough to have differentiated, or which were subsequently smashed apart, are known as non-differentiated bodies.

Saturn has definitely differentiated. Its core is made of metals, the same as the Earth, and there's likely to be a rocky core surrounding it, also like the Earth. But from there on out-wards is a tremendous swirl of gas, almost entirely hydrogen and helium, two of the lightest materials in the Universe. And because *so much* of Saturn is made of extremely light materials, on average, the whole planet can still be less dense than water, even though if you were to randomly sample chunks of Saturn, you'd have very poor odds of pulling something with the same density as water.

But ignoring that minor detail, as far as averages go, it's a straightforward calculation. Planets are spheres, and we know how massive they are, and how big their sphere is, so you can pretty easily figure out their density. NASA has a handy online chart listing the densities of all the planets (plus the Moon and Pluto), which, in kilograms per cubic meter, range from a high of 5,247 (Mercury) down to 687 (Saturn).

Water is 1,000 kg per cubic meter. Saturn is the only planet under this threshold – the next lowest density planet is Uranus, at 1,271. So we arrive at our conclusion that Saturn is less dense than water!

On Earth, if you have small objects which are less dense than water, they will float. Things like tennis balls, cans of diet soda, and surfboards will all sit on the surface of water.

This works because the density difference between the object and the water it's sitting in creates a buoyancy force which is stronger than the force of gravity pulling downwards. If you're a scuba diver, one of the goals is to become 'neu-trally buoyant' – the point at which those two forces exactly equal each other, and you won't rise or sink without your own control.

Saturn being less dense than water, unfortunately, does not mean it would float in a bathtub, if you could magic a planetary-sized bathtub into existence, because the gravity part of this balancing act starts to become really important. Saturn has *a lot* of mass, and so its own gravitational pull is quite strong. Not true of the surfboard on Earth's oceans, which is tiny compared to the mass of the planet.

If you want Saturn to float, you'd need a bathtub so massive that Saturn's mass is small by comparison.

So now we're talking a giant sphere of water, which has its own tremendous gravity, such that Saturn's own 10^{26} kilograms of material is small.

Let's say it's 10^{30} kg, which gives us a factor of 1,000 difference in mass from Saturn.

Congratulations – you have made: The Sun. The Sun has 10^{30} kg-worth of mass, and water is a denser material than the hydrogen the Sun is made of, so our water sphere is probably going to *immediately* start nuclear fusion. Our Extremely Large Bathtub will not be water for very long, but will instead become a plasma of hydrogen and oxygen in which approximately *zero* things will float because they will all be incinerated.

Small problems have been encountered. Let's not try this particular experiment.

If, on the other hand, we had a *small* object with the same density as Saturn, that would float, while small proportionally dense models of all the other planets would sink to the bottom of whatever body of water you'd put them in: a much safer version of this experiment.

VENUS'S SURFACE IS NEW

Welcome to Venus! It's a hellscape.[*]

On your way in, do mind the 50 km-thick cloud layer, sulfuric acid rain, blistering heat, and crushing air pressure at the surface.

There's also an ongoing forensic examination, because Venus has a serious Case of The Missing Craters. The culprit: definitely Venus. The mechanism: definitely volcanoes. After that: things get a little less clear.

Craters are a pretty reliable feature of rocky bodies in our solar system; as examples, we can look at Mercury and the Moon, which are positively covered in craters. Even Mars is fairly pockmarked. Initial conclusion of our forensic inspection: these rocky bodies regularly get hit by smaller space rocks.

If you look a bit more carefully, it's clear that most of the time, the biggest craters on all these surfaces are themselves covered in smaller craters. Which then tells us the biggest craters were there first; if the big crater had happened second it would have obliterated any smaller craters. This 'big craters are early' is a common feature across most of the rocky bodies with lots of craters in our solar system, and from it we concluded that Big Rocks That Make Big Craters were common a long time ago, and are less common now (life on Earth gratefully

---

[*] As long as you don't ascribe to Dante's cold version of hell.

pays its respects). Small Rocks That Make Smaller Craters are still hanging around in the solar system, and they pummel surfaces relatively indiscriminately. Periods of frequent impacts in the younger solar system are known as 'bombardment', and come in two theorized flavors, 'Early' and 'Late Heavy'; the exact relative strength of the interplanetary game of dodgeball that gravity was playing is still to be determined.

So how can Venus be so lacking in craters? Compared to Mercury, its surface is positively *glossy*.

It's incredibly unlikely that Venus's surface is so smooth because it never got hit by any space rocks; even the Earth we know for a fact has been hit more than a few times. So instead scientists have turned their heads to a different question: How do you erase almost all the craters from the planetary surface? Whatever the mechanism, it means one thing for sure: If you're erasing craters, it's because *something* is happening on the surface, and that means the surface is active — laid down more recently than the craters we would have seen if nothing had happened to erase them.

The Earth is another world without quite so many craters. And that's because the Earth has an active surface recycling program, through a combination of plate tectonics, which destroys the crust on which a crater might have existed, erosion, which erases visible signs of craters on the surface, and volcanism, which just lays down more rock on top of whatever was there before.* So perhaps Venus is stealing from our bag of tricks, and is using one of these to get rid of its craters.

. . . .

* Pompeii can attest that sometimes this process is a bit abrupt.

Venusian curveball number 1: There are no *small* craters (smaller than 3 km across) on Venus either. This one we know the answer to. Remember that 50 km blanket of crushing, toxic atmosphere?* That atmosphere is so thick that the smallest objects encountering Venus will simply burn up in it, the way that meteors do in the Earth's atmosphere. Venus can destroy larger rocks than the Earth can, because the atmosphere is denser and so it is able to wear these rocks down into nothingness. However, even this wall of dense carbon dioxide won't stop a large enough rock. So the atmosphere can stop small things, but this doesn't explain the lack of Large Craters.

Venusian curveball number 2: The craters which are there are really crisp. This functionally excludes gradual processes like erosion on Earth. It also excludes any way of getting rid of craters that happens very slowly, over long periods of time, because then we'd expect to see a lot of half craters, eroded craters barely visible, instead of what we have, which are Hot Freshly Baked Craters.

So what's left of Earth's tricks? Plate tectonics and volcanoes. On Earth, these two are connected. Venus, by contrast, doesn't have plate tectonics the way that the Earth does; but it's still geologically active: it is absolutely covered in volcanoes.

So we're back to where we started: the culprit is definitely volcanoes.

But *how* is it volcanoes?

---

* Venus really isn't a great holiday destination.

If there's one thing volcanoes are famous for, it's 'creating a new rocky surface around them', and so on the list of things that can erase a crater, they're a good bet. But now we have a new question: Are there enough volcanoes, and are they active enough to have erased *all* the big craters? On Earth, volcanoes erupt, cover the area around them, and (mostly) subside again for some time. Venus doesn't play by these rules. Venusian volcanoes are a different scale of geological feature entirely. Not only do they cover a much larger surface area than the typical volcano on Earth, but there are volcanic features on Venus which just don't occur on our planet. This is probably due to the differences in surface pressure, temperature, and the profound dryness of Venus.

So let's say we'd like to cover up the planet with lava, courtesy of Venus's volcano population. We basically have two options. Either we do this 1) all at once, or 2) over time, randomly.

We have one further tricky thing to work into our solution: however you wipe out the oldest craters, you have to leave a random assortment of craters to be visible to human-built robotic craft in the present day. And by random, we really do mean random: there is absolutely zero pattern to where the craters are on the planet.

So one option is that Venus's volcanoes have scattered themselves pretty randomly over the surface of this hot mess of a planet, and as they pop up, they wipe out the craters beneath them, and subside. Other volcanoes will arise in other areas, wipe out the craters in their general area, and repeat for several billion years.

The other option, 'cover up all the craters at once', is dramatic, because it means we're going to make nearly the entire planet a lava ball for a little while.* This theory is dubbed the 'catastrophic resurfacing event',† and states that about 400 million years ago (give or take a few million years), almost the entire surface of the planet was covered in freshly laid basalt (aka very hot lava).‡ This would, indubitably, wipe out any trace of the craters which might have been there. This event isn't thought to have lasted very long (astrophysically speaking) – only about 10 million years or so – and after that, things could cool down enough to stop being a planetary-scale supervolcano. New craters would punch into the surface, and the level of volcanic activity since then could be much less, explaining why the craters we see now are so clean-looking!

Both options have valid objections. If the steady state option is correct, it's hard to explain how the craters that are there are all pretty much untouched by volcanism, and almost exactly randomly distributed, so it seems that no part of the Venusian surface is younger than another. If some portions of the surface were much younger than others, we'd expect to see empty gaps where craters are *not*, because the fresher land there

---

* Temporarily rendering Venus the solar system champion of The Floor Is Lava.

† When astronomers use the word 'catastrophic', it's no hyperbole. Let's just melt the whole planet into lava.

‡ Paul Byrne, a planetary scientist, described it this way: 'We're talking about an affront to God in terms of the amount of lava that comes out per unit time.'

would have covered any old craters. Instead, no part of Venus has a tell-tale missing crater patch.

If we go for the catastrophic model, then we have to be able to both start and stop a planet-wide tectonic melt on a relatively short time-frame, at a scale which has never been seen on Earth. Currently, the hypothesis runs that the core of the planet overheated due to radioactive decay (this also provides heat to the core of the planet on Earth) but instead of being able to gradually vent heat through plate tectonic activity, Venus made a hotspot (like the one underneath Hawaii) but which covered a massive fraction of the entire planet.

It is, of course, possible that the answer to this 'which' question is 'yes', and that aspects of both models have been in play over the course of Venusian history.

It would help if the best map of Venus's surface was more recent than *1994*. The highest-resolution images that we have were made by Magellan, a robotic spacecraft launched in 1989, arriving at Venus in 1990, and which then spent four years in orbit echo-locating the surface through the clouds with radar. Bounce radio waves off the surface, wait for them to come back; the signal that reaches you first hit a taller object. At the end of its mission, Magellan plunged itself into incineration in the Venusian atmosphere. It's been almost 30 years, and these are still the best global maps of Venus that we have. But we finally have new missions scheduled, so both the advent of better maps and the changes in the surface features between 1994 and today may help us to understand how Venus works in much more detail than we have been able to do so far.

No matter what explanation planetary scientists settle on, it's got to involve a way to erase old craters, leaving new ones intact, and it's going to involve a medium to catastrophic volume of lava. But hey: shiny new Venusian surface!

THE MOON'S WET

With daytime temperatures way past the boiling point, one would not expect the Moon to be able to host any kind of water. Ice should evaporate, liquid water should evaporate even faster, and water vapor — as a gas — should just drift off into outer space.

And yet, somehow, the Moon's sunlit surface* seems to be able to keep some water around in the form of ice.

Ice on the Moon isn't a new finding. There are some deep dark craters, especially towards the poles, that should be in permanent night. Any kind of divot near the poles creates a hiding place the Sun will never reach; very convenient if your goal is to stay very, very cold.† The temperatures in them can stay colder than any freezer on Earth, at −210 C, and in these spaces, ice can hide unperturbed. These locations are called 'permanent cold traps', and we've detected water in them going back to 2009, when an Indian spacecraft, Chandrayaan-1, uncovered the signature of water from orbit, and a NASA spacecraft called

....

* Reminder that all parts of the Moon will be sunlit at some point in the Moon's orbit around the Earth. Aside from the permanently shadowed craters near the poles, there's no permanent 'dark side' of the Moon, though we only ever see one side from Earth.

† And if your goal is to never see the Sun, they're equally convenient. (Are you a vampire?)

LCROSS smashed itself into the Moon – among the debris it kicked up was water.

But away from the poles, where the blazing Sun shines for two weeks at a time, it's been long expected that there shouldn't be any water. The Sun has been roasting a surface that doesn't provide much shelter for some 4.5 billion years – any water that might have been there in the past should certainly have evaporated long ago.

Except that apparently it hasn't.

The SOFIA telescope, which itself is a feat of equal parts impressive and hilarious human ingenuity,* detected the signature of water away from the poles of the Moon, exactly where it shouldn't be, according to all logic and expectations so far.

How exactly is water, a known Thing That Evaporates In Even Moderate Warmth, still on the sunlit surface of the Moon, which reaches over 120 degrees Celsius in the sunlight?

There are really only two options for how this is possible. First, the water could be trapped inside the rocks on the lunar surface, and so it's not evaporating because it's stuck inside a rock, but still detectable as a material near the surface. Or the water could be actual ice on the actual surface, and somehow not completely fried away just yet. Both options could work,

---

* SOFIA is a telescope that observes in a wavelength mostly blocked by our atmosphere. So we strap a telescope to the insides of an airplane; then we fly that plane as high as we can manage, and just open it up to let the telescope see out, now that it's above the densest part of the atmosphere, which is near the ground.

but to distinguish between them, we'd have to take a more careful look, because there should be subtle differences. The most obvious of these is that the 'embedded in rocks' explanation should probably not come with any changes as a function of time. If the water is trapped, it's trapped; waiting for a few days shouldn't change anything.

But in our observations we found that there was more water detectable in the afternoons, when the sunlight is less intense, and less water in the lunar morning, when the sunlight is most direct.

This is exactly the kind of change as a function of time that would be impossible to explain if the water is stuck inside the rocks, so we're back to trying to figure out how water can exist on the surface without totally evaporating.

It seems that the answer, like the devil, is hiding in the details.

Generally speaking, when people simulate what should happen to water on the Moon, they model a smooth, featureless surface. And broadly speaking, this is true. There's no architecture, or foliage, or really any large topographic changes, so it wasn't an unreasonable thing to start with.

But we know, both from high-resolution imagery and from the Apollo astronauts who visited, that the details of the lunar surface are not so polished. There are large (compared to a human) boulders, and while the surface of the Moon is covered in fine powder, there are also a lot of rocks of all sizes. And these rocks of many sizes, while they don't make for *large* features, do make for a change, relative to a perfectly smooth, billiard-ball Moon.

Because the Sun is rarely perfectly straight overhead, each little rock casts its own shadow, and the shadows on the Moon are stark in a way that Earth shadows cannot be, because on Earth we have a thick atmosphere. Even with our atmosphere, though, the temperature in the shade on Earth can be dramatically lower than the temperature in the Sun, especially in desert climates, and that's *with* air moving around.

On the Moon, the only thing that can move heat around is the rock itself, which is much less efficient at that job. And so the question we need to ask, in order to answer 'How hot is this specific surface?', is: 'Are you in *direct sunlight?*', which is a different question from 'Are you on the sunlit half of the Moon at this moment?' A study found that you could have cold traps of permanent shade only centimeters across – any smaller than that and the rocks do a good enough job of transferring heat that it would warm up the shadowed regions.

A follow-up paper to the SOFIA detection looked at exactly how the presence of rocks and their shadows might allow more water to stay on the surface of the Moon, and it concluded that the shadows of even the smallest rocks work as cold traps; as the Sun slowly progresses across the lunar sky, the shadows the rocks cast will shift. The newly exposed frost can and does evaporate at that point, but because there are still cold traps nearby, as long as the evaporated water isn't unlucky, it can settle back down – freezing out again – in the slightly moved but still nearby shadowed area.

It's worth noting that this is not a lot of frost. The Sahara desert is still 100 times wetter than the Moon, and so if you're looking for more substantial stockpiles of water, those shadowed craters in the polar regions are probably a better bet. And indeed, the volume of water at the poles (particularly the south pole) should be much larger than the amount that can persist closer to the equator.

Learning how and where the water sits will be really useful if we want to set up a longer human habitation on the Moon. If there's water only at the poles, then any habitat established further away from those areas would have to bring *all* of its own water, and water is notoriously heavy.

If the Moon has water – even smaller amounts – away from those polar regions, that could be convenient. A lunar base near the poles could have some problems with power; because the sunlight is always *so* oblique, solar power arrays might struggle. Solar power is the most practical long-term way of powering any kind of habitat, since it is not going to run out of supply during the two-week-long lunar day. You would, of course, need a different strategy for the equally two-week-long lunar night, but it would be nice not to have a power supply problem 100% of the time; setting up camp further from the poles would solve at least that problem.

There's a period of time, in this picture of the rock shadows acting as cold traps, where the Moon has some hovering gaseous water – not much, but a tiny bit. And if you have the thinnest possible haze of gas surrounding your rock, while you haven't technically graduated to having an 'atmosphere', we do get to name the gas 'haze', and it's been dubbed an 'exosphere' to distinguish it, because it is tremendously thin.

It's about 10 trillion times less dense than Earth's atmosphere at sea level – for all purposes it's 'a vacuum'. It's about as dense as the Earth's atmosphere is where the ISS orbits, and the astronauts there are very invested in not being exposed to that flimsy edge of our atmosphere.

SOME OF TITAN'S LAKES MIGHT BE THE FLOODED REMAINS OF EXPLOSIONS

Titan is a bizarre moon. It orbits Saturn, has a thick, opaque atmosphere, and is way too far from the Sun to host liquid water.

And yet, it has an active liquid cycle, with rain, clouds, rivers and lakes. In fact, it is the only place in the solar system, aside from Earth, with lakes. It's just that the liquid flowing is methane. Methane on Earth is a gas (it catches fire, and is the flammable component of natural gas), but Titan is so far from the Sun and so cold (−179 C) that methane functions in many of the same ways that water does on Earth.*

The Cassini spacecraft swung past Titan many times; and it is from Cassini that we have gathered most of our information. Cassini also carried a landing probe to drop onto Titan; the Huygens probe descended safely onto the surface in 2005, broadcasting images as it descended.

But the most transformative information we learned about Titan was from radar mapping the surface as Cassini swung past it, and from watching changes over time in what those maps revealed.

. . . .

* The methane on Titan would *not* catch fire if you brought a match; that kind of burning requires oxygen, which Titan doesn't have – its atmosphere is mostly nitrogen, mixed with methane.

Radar is a pretty simple concept; you just shoot radio waves at an object, and wait to see both *how long* they take to bounce back at you, and *how much* of that radio light is bounced back.

Simply counting how long it takes to bounce back tells you how far away the object is. As it applies to the surface of a moon, this means that the signal you receive back first is from the part of the ground that is most elevated, and the signal you receive last will be from the lowest depressions below you. This process is known as radar altimetry, and has mapped a lot of objects in the solar system.

Then there's the 'brightness' of the surface. While brightness in the visible range is pretty easy to understand, 'bright' to radar is really a measure of reflectivity, and there's more than one way to be reflective. One way is to have an irregular surface; very flat surfaces tend to absorb a lot of radio light.

So, if an object is both irregular in texture and highly elevated over the surrounding region, you'll both receive a lot of signal back, and it will come back quickly. If you have a very smooth surface in a low-lying region, you'll get very little back, and whatever you do get back will take a longer time to return.

On one of the earlier radar passes of Titan, folks noticed that there were irregular, blob-shaped, *very* dark regions on the surface. This meant the regions had to be very flat, either because they were pools of liquid or because they were made of very fine particles. The easiest explanation came from examining the areas around them: many of these dark patches had very river-like, also very dark, features feeding into them. Liquid was the simplest explanation. These regions were dark because they were flat lakes of liquid methane.

In looking at these lakes more carefully, it became clear that there were a few things going on. Some of the lakes looked like they were partly evaporated. Others looked like they were as full as they ever get. Some of the lakes were large – 1,000 square kilometers in size, often with clear rivers to feed them. Others were very abrupt divots, with no discernible river fueling the lake itself; these earned themselves the name 'sharp-edged depressions'.

These sharp-edged depressions came with another surprising feature: many of them have a raised ridge, the most extreme rising 300 meters tall, surrounding the lake. (More typically, the ridges were 'only' 100 meters high.)

Raised rims don't generally appear if you have a lake that appeared through any kind of erosive process, because the first thing that the erosion would do is get rid of that ridge! And it's not just a few lakes with raised edges; it's *most* of the smaller lakes. 75% of Titan's small lakes seem to have fortressed walls, somehow.

While you don't get raised walls with erosion, you absolutely *do* get raised edges if what you're looking at is the remains of some explosion. Now, with an explosion, you have two options. The first is to have something slam into the surface of your moon (also known as an impact crater). If you punch a hole in the surface that way, then the liquid methane could find its way into this depression and create a lake. Flooded craters*

---

* Though more often they're flooded with lava, rather than any kind of liquid.

have a long and glorious history in the solar system, so finding them on Titan wouldn't be a surprise.

Except of course that impact craters tend to be round, and most of these ones aren't. But some kind of thing must have happened to create a 100-meter-high rim around the lake. So onwards to the second option: maybe the ground just exploded.

That may sound silly, but it's currently a working theory.

And it's not that far-fetched, because the ground has exploded on Earth occasionally too. If you have lava rising from deep underground, and it encounters a substantial volume of water, the lava will rapidly boil that water, and converting water to steam in double-quick time tends to cause a rather large explosion, as the water vapor would like to expand, except that it's trapped under rock, very inconveniently. And so whenever the pressure of the water vapor exceeds the rock's ability to hold itself together, you explode the ground, and create what's known as a maar.

Because you just flung the ground into the air, the displaced ground is going to come back down again,* and so it tends to pile up around the explosive hole you just made, rather than going particularly far away, making an elevated rim.

For Titan, the idea runs like this: Once upon a time in the distant past, Titan was even colder than it is now, and instead of having lots of liquid methane around, it might have had *nitrogen* lakes. (Nitrogen requires even colder temperatures to stay in liquid form, and Titan is now too warm for such lakes.)

....

* Gravity takes no days off.

If your surface is full of nitrogen lakes, then you could equally have little aquifers of nitrogen under the surface.

Then we let Titan warm up, and at some point there's a temperature at which the nitrogen goes from being a liquid to a gas. If there are pockets of nitrogen underground, then the expansion the liquid undergoes as it transitions to a gas has to find some way out. If there's no easy way, then the pressure will build up, as it did for the water vapor that created the maars on Earth, and eventually the nitrogen is just going to catastrophically burst out of the ground.

Lo and behold: the ground has exploded.

Now that you have a cavity from your exploding ground, and a warmer Titan that permits liquid methane, you can just fill in the divot you made with methane, and presto. You've got yourself a methane lake with a raised rim.

As far as a filling mechanism goes, you could fill it with the methane rain that we've observed falling on Titan, or perhaps a methane storm, something we've seen trigger flash-flooding-style rainfall onto the surface of Titan. But however you fill your lake, it seems you have to build it first.

We should be learning a lot more about Titan's geological processes in the next fifteen years, because we're sending a quadcopter there. The atmosphere is so dense, and yet not very windy, that we can send a robotic science craft on helicopter blades to hop its way around the moon, relaying information back to Earth after each flight. The spacecraft is called Dragonfly, and it is set to launch in 2026 and land in 2034.

PLUTO'S SURFACE IS
YOUNG, SOMEHOW

Before New Horizons flew past Pluto in 2015, our best guess of what it looked like was based on very low-resolution images from the Hubble Space Telescope, which had hinted at some variation in color across its surface, but precious little past that. We assumed, at the time, that it was a slightly mottled ball of ice and potentially some rock, and very likely covered in craters, as other things would have continuously hit it since the formation of the solar system.

Craters, as we've mentioned with Venus, are one of the easiest ways to age-date a surface. Given some cratering rate, if we spot a certain number of craters, we can estimate how long that surface has been there as a landing zone for any variety of rocks that might have whizzed into it.

But when New Horizons went cruising past Pluto, it sent back images of a remarkably varied little world, and among its surprises was the *exceptionally* crater-free Sputnik Planitia, which you might know as the 'heart-shaped region' on Pluto. There were a grand total of *zero* craters that could be distinguished in the images of Sputnik Planitia.

We have to be slightly careful with wording here, because it's certainly possible that there are *some* craters on Pluto's heart-shaped plains, but down to the limits of the crispness of our images, there were none — certainly there aren't any *large* craters on Sputnik Planitia. But small ones? While New Horizons was going past, it managed to take both low-resolution images (of a large area) and some high-resolution imaging (finer details, but in exchange, less of Pluto).

There were no craters in either of them. From the low-resolution images, we know there can't be any craters bigger than 2 km across anywhere on Sputnik Planitia, and the swath imaged with the higher-resolution camera means that in that region, there are no craters bigger than 625 meters. That's pushing down into Very Small Crater territory, on a cosmic scale. Meteor Crater, in Arizona, is bigger than this second limit, at 1,300 meters across, and it's a very recent crater – only 50,000 years old at most.

Before New Horizons went zipping by Pluto, the prediction was that there should be somewhere between 40 craters and 50,000 craters larger than 30 km on the surface, built up over the course of the last 4 billion years, if the surface is old and has retained the record of all its craters. Even before arrival, scientists were working out what it would mean if they saw fewer than the expected number, because we knew there was nitrogen ice on the surface, which behaves very differently than regular old rock. One modeling attempt found that the nitrogen could just evaporate away into Pluto's atmosphere, which would erase shallow craters. Nitrogen ice, at the temperature of Pluto, can also act like a very thick fluid, and so some craters might also just slump flat, which would make them harder to spot. In either event, these were just modifications of an expectation to find a *lot* of craters on Pluto's surface.

Sputnik Planitia accounts for about 5% of Pluto's surface

area, so there should have been somewhere between two and 2,500 craters on it. But instead there were none. Whatever bigger craters might have existed on this bit of Pluto at some point in the past, they're gone. Much like the glossy-smooth surface of Venus, we must conclude that Sputnik Planitia has been freshly laid down, and has covered up the evidence of any previous pummeling. Purely based on the lack of craters, Sputnik Planitia had to be no older than 10 million years.

Pluto's surface isn't universally baby-faced; just to the west of this heart-shaped region is a darker patch of ground informally known as Cthulhu Macula.* This area of Pluto is indeed full of craters, the way you'd expect of a 4 billion year-old surface. Over the entire planet, about a thousand craters have now been mapped, which is a small number for something the size of Pluto. But Sputnik Planitia isn't contributing to the count at all; somehow, the heart-shaped region is newer than the rest.

Pluto has none of Venus's volcanoes, as it's drifting at the edges of the solar system and very, *very* cold. So how can Pluto have erased its craters? The current best guess is a strange feature of nitrogen at the very cold temperatures of the outer solar system. Nitrogen, on Earth, is a gas, and even getting it to be a *liquid* requires extremely cold temperatures.† It begins to boil at −196 C (77 Kelvin), and only begins to freeze once you get down to an even more frostbitten −210 C (63 K). Pluto's

---

* Yes, astronomers are nerds. Charon, Pluto's moon, has a region named Mordor Macula.

† It does make delicious ice cream, though.

surface temperature normally hangs out somewhere around the −238 C (35 K) mark, *easily* cold enough to have nitrogen ice.

If you look carefully at images of Sputnik Planitia, you can see what looks like a large number of blob-shaped regions, each surrounded by a tiny moat. These, we think, are the signature of a very slow, very cold boil of the nitrogen ice on this part of Pluto, and are technically called 'convection cells'. Convection is a general term for the movement of warmer material upwards – the most familiar form is that of a rolling boil, but it can take place in any kind of medium. In the case of Pluto, underneath the extremely frozen surface, things must be a little warmer, and this is enough to make the nitrogen act a little less like ice and a little more like a fluid, and it can slowly bubble up in the centers of these cells, allowing the colder ice to sink back down in the troughs between cells. Not only does this explain the unique shapes of these blobs, but this convection is an *excellent* way of removing craters from the general area.

As result of this convection, we suspect that the surface itself is quite young! Estimates on the depth of this pile of bubbling nitrogen slush vary, from 10 km thick to perhaps only 4 km deep. It's likely to be filling in an ancient impact crater that we can no longer see, because it's turned into a giant cauldron of nitrogen. The age of Sputnik Planitia is slightly dependent on how deep you think the ice is, but now that we've been able to spend some time modeling how this region might actually be behaving, all the models agree that it can't be more than a million years old, and perhaps only 180,000 years old. The coelacanth is 400 times older than this part of Pluto's surface, even with a surface a million years old.

Sputnik Planitia's slow bubbles aren't the only sign of very new surfaces on Pluto – there are many more, but one of the more fun is that it also has glaciers, made of the same nitrogen ice that forms the majority of the bubble zone. At the temperatures of Pluto, water ice is effectively like rock, but nitrogen ice is able to mash itself around and be a little more flexible than water ice on Earth. And water ice on Earth is what all of our glaciers are made of, so these nitrogen glaciers will be, if anything, even more fluid than our own.

We can see glaciers from orbit around Earth, and they have a very tell-tale texture to them. You wind up with wrinkles and folds which point along the direction the glacier is flowing, looking for all the world like a river frozen in time (and in this case also literally frozen). We can see these same shapes on Pluto, even though they'll be made out of nitrogen instead of water ice. They're very common around the edges of Pluto's heart, which is itself a bit of a low-lying area, so as these glaciers flow downhill through valleys and rifts, they eventually reach the equally fresh ground of Sputnik Planitia.

All of these fresh, clean surfaces require a little more heat to fuel than we expected Pluto to have; part of the puzzle that remains is still the question of 'How is Pluto not frozen solid?' Something must have warmed it since the solar system formed, and what exactly is still a mystery.

SOME ASTEROIDS ARE JUST
PILES OF RUBBLE IN SPACE

When most of us think of an asteroid, we think of a large, rocky or metallic chunk of material, hanging out in space. If we're particularly keen on the dinosaurs, we might also start thinking about asteroid impacts with Earth. The dinosaur era was brought to a geologically abrupt* halt with the introduction of a ten-mile-wide asteroid to the Yucatán peninsula.

Asteroids are the left-over nuggets of material that didn't make it into a planet during the formation of the solar system, so if we have small chunks of rock or metal, we can pretty reasonably assume that they would have formed into solid, though not necessarily round, objects. And indeed, we do see a range of lumpy, potato-shaped objects in a survey of the asteroid belt.

We humans haven't sent all that many robotic craft to go visit asteroids, and even fewer to get up close and personal. That said, there have been a few. Twelve in total, as of 2020, of which only four have spent more time than just zipping past on their way to somewhere else. NASA's Dawn mission was

. . . .

* For anything near the impact crater, the era was brought to an abrupt halt by any metric; 20 seconds after impact, simulations predict a 20 km-high molten rock tsunami, excavating a hole 30 km deep and 60 km wide. Now, I'm no biologist, but I'm pretty sure that has a 0% survival rate.

an orbiter around two asteroids, Vesta and Ceres, the latter of which is the largest object in the main asteroid belt. There have been two Hayabusa missions operated by the Japanese space agency JAXA, each of which returned to Earth with a sample of the asteroid they visited. And there is OSIRIS-REx,* NASA's version of the same – with the mission of returning to Earth with a piece of asteroid it collected.

The three asteroids we've visited for sample return, Itokawa, visited by Hayabusa 1, Ryugu,† visited by Hayabusa 2, and Bennu, visited by OSIRIS-REx, have all been fascinating objects, and they've all had something in common – they're the shattered remains of some other, probably larger, object.

Dubbed 'rubble piles', none of these three have the density you'd expect from solid rock. In fact the only way you get to the low densities they've been measured to have, is by having substantial cavities of nothingness in the middle of your pile of rock. So not only are these three asteroids not a single rock, but they're not even a particularly *well assembled* pile of rubble. It's a bit of a packing materials model: take up a lot of space, but have a lot of gaps between each object. You could, technically, describe them this way: 'an unorganized collection of

---

* This is one of astronomy's Famous Tongue Twister Acronyms: 'Origins, Spectral Interpretation, Resource Identification, Security, Regolith Explorer'.

† Ryugu is named after the underwater castle of a dragon in Japanese mythology; a fisherman visited the castle for three days and returned home to find that 300 years had passed.

macroscopic particles (rubble) held together by their self gravity', which is just technical speak for 'a messy jumble of rocks of all sizes, which only hang out together because they feel the gravity of all of their assembled pebbles'.

On Earth, most rocks have a density of around 2.5 grams per cubic centimeter. Ryugu, by contrast, has a density of only about 1.2 grams per cubic centimeter, which, considering it is still made of rock, is an impressive *half* the density of typical rock. The only way you get there is if your object is about half empty void.

There are two places for empty voids: *between* the boulders your objects are made of, or *inside* the boulders your objects are made of. Very much like styrofoam. You can pack loose pieces of styrofoam quite densely, but because the styrofoam itself is also full of air, the overall density is low. Alternatively, you could make slightly more dense packing material, but pack it very loosely, and get the same low density. A more careful study of Ryugu's composition indicated that while, overall, it is about 58% empty, only about 16% of that is spaces between boulders; the rest is just bubbly rocks.

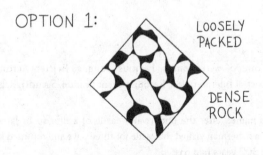

OPTION 1:

LOOSELY PACKED

DENSE ROCK

OPTION 2:

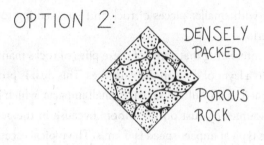

DENSELY PACKED

POROUS ROCK

Rubble pile asteroids aren't that uncommon – in fact, for asteroids between 200 meters and 10 kilometers across, it's thought that rubble piles are the most common format. From this we learn that the asteroid belt is fond of smashing its pieces together. Rubble piles are formed when two objects collide sufficiently vigorously that both are shattered into tiny fragments. Gravity will pull the bouldery mess at the end back together, but there's no longer a single, solid rocky object; it's simply a cosmic pile of Mixed Boulders, held gently together by gravity.

Bennu is so comprehensively a pile of boulders that the entire surface is composed of robot-smashing, jagged corners. It took the science team nearly a year to find a safe place for OSIRIS-REx to do its sample acquisition. The surface of these rubble pile rocks may be slightly more rocky than the interior, due to a phenomenon known as the Brazil Nut Effect (yes, seriously).* That is, when you shake a tin of mixed nuts, all the Brazil nuts – the largest of the nuts – rise to the top. Since your rubble pile asteroid is just Mixed Boulders, the same thing can happen if you rattle it; some of the boulders may shift to the

---

* Okay, it also goes by the name of Granular Convection.

surface, with smaller pieces of rock and dust settling towards the interior.

It's interesting that some of these piles of rocks manage to maintain a layer of dust on their surfaces. This dust is probably not formed as a result of repeated small impacts, which is one of the sources of dust on the Moon, because in the asteroid belt, the typical impact speed is 5 km/s. The typical escape velocity of a rubble pile asteroid is only 1 km/s; so if you thwack one of these, all the pieces of grit you would let loose from a larger boulder would just twang out into space, never to be seen again.

This dust may come from an odd place. A lot of these asteroids spin pretty quickly: Bennu spins on its own axis once every 4.3 hours, with Ryugu only a little bit slower at once every 7.6 hours, and these are not considered particularly rapid rotators. Going from sunlight to the dark void of space every few hours creates a temperature stress on the rocks; swelling slightly in the heat of the Sun, and shrinking down again in the cold, repeatedly, every few hours, for over 4 billion years. Over time, this could crumble boulders into dust.

Because they spin so quickly, many of these piles of rocks also assemble themselves into roughly the shape of a top. Ryugu and Bennu have acquired this shape: diamond-shaped as seen from the side, round as seen from above or below.

Itokawa, on the other hand, looks like a peanut shell. It rotates more sedately, at once every twelve hours, and its two halves, in addition to both being rubble, have different densities. This led scientists to suggest that Itokawa, in addition to being a pile of rocks held vaguely together by gravity, was once two piles of rocks held together by their respective gravities, until they came together, very gently; and instead of bouncing, their mutual gravity allowed them to pull together. This type of object is called a 'contact binary' – a set of what used to be two objects just ever so gently becoming a single one.

Contact binaries are relatively rare overall, but common in the rubble pile population.* In general, things in space are moving faster, and they have more of a smash and/or dash approach to interacting with each other. But if your asteroid is a rubble pile in space, contact binaries are a little easier to form, since each object is only tenuously held together in the first place. When gravity pulls the two objects together, their accumulated boulders will shift and adjust near the contact point.

Why do we care to study these things? One reason is 'because it's neat'. The second is much more selfish, but extremely practical.

There are a large number of asteroids which have orbits around the Sun that cross over the orbit of the Earth. Generally the timing works out so that when they cross our orbit, the Earth is not there, and no harm is done. But eventually, on potentially long timescales† of 'eventually', there is bound to be some asteroid with an orbit that crosses the Earth's path while we are in the way. This is not generally considered a good day.

---

* Somewhere around 14% of near-Earth asteroids are thought to be contact binaries.

† Though this depends on having an accurate census of Earth-crossing asteroids, which we don't.

So, we have lofty goals of being able to nudge asteroids out of the way of the Earth. We don't need to do much; just slowing down or speeding up the asteroid a tiny bit is enough of a delay to let the Earth slide though unscathed. But that kind of nudging works best if you have a solid object to work with. Rubble piles are hard to nudge, because — well, they're just a pile of boulders. How do you nudge thousands of kilograms of boulders, collectively? Not easily, since one of the defining characteristics of a pile of rubble is that it's not held together very well, and is easily disrupted. If you pull one near a larger object, it should break apart into the zillion pieces it's made of.

Studying them close up will let us understand them, and their structure, a little better. We can hope that we will not need to nudge a rubble pile asteroid, but it's best not to go in unprepared.

JUPITER'S MAGNETIC FIELD WILL SHORT-CIRCUIT YOUR SPACECRAFT, BUT VENUS WILL JUST MELT IT

When we imagine sending spacecraft to the other planets in our solar system, there are a set of hazards we might immediately think of. You'll need some source of power – the further out you get from the Sun, the worse an idea solar power is – and you'll need the electronics you carry with you to be robust in the cold temperatures of outer space, but a problem that you may not have thought about is that some parts of space are very much a gigantic particle accelerator, and if it sounds like a bad idea to put your laptop in the Large Hadron Collider,* you would be *absolutely correct*.

The problem (and benefit) of living on a planet with a thick atmosphere and a sturdy magnetic field is that we can ignore a lot of radiation from the Sun and other places. This is nice, because radiation is usually bad news, but it also means that once we start equipping electronics to operate anywhere beyond the cushy surface of our planet, we have to start building in a *lot* of extra precautions.

Even the ISS, in low Earth orbit, and still well within the

---

* Experiments we don't need to actually conduct, for $2,000. We can confirm, however, that weasels chewing on power cables will both shut down the LHC and be thoroughly fried to a crisp. Sad experiments all around.

magnetic field of the Earth, has to have extra protections in place against radiation, and it's still a pretty protected space, by the solar system's standards. Our delicate human bodies don't tend to do well with *any* extra radiation, and radiation sickness is a nasty business. So the standard for how much radiation the ISS should expose people to is 'ALARA' – 'as low as reasonably achievable'.

As we go further and further out, more and more shielding would be required to protect a human – but robots generally have a higher tolerance to Radiative Nonsense than humans do.

Not an infinite tolerance, though. If you bombard any computer with enough high-energy pellets of stuff, one of two things can happen. You can gradually build up a static charge, which could eventually discharge and cause a short. Or, you can bash the computer abruptly, which can have the effect of scrambling it by changing too many 0s to 1s.

Building up a static charge can be mostly avoided if you try not to build your spacecraft with tactically placed conductive materials, which means the charge can discharge outwards towards space, rather than across the spacecraft (which fries it), as many a lost spacecraft can attest.

But the more irregular bit-punching by unfortunately aimed small pieces of atom is harder to guard against, and the more intense a radiation field you're in, the more you have to work to keep your computers functional.

Welcome to Jupiter. The magnetic field here is 10,000 times stronger than Earth's. Its radiation environment is known, in technical circles, as 'intense'. Much as the Earth creates a donut of high-energy particles when small particles get caught up in its magnetic field,* Jupiter can do the same, but 10,000 times more. We try to avoid sending things to live in the Earth's donut of high-energy particles, and Jupiter's belt is even worse.

So, if you'd like to send a spacecraft anywhere *near* Jupiter, and have it survive to tell you about it, this translates into building heavy, heavy shielding for the computers on board. The Juno spacecraft, which arrived at Jupiter in 2016, has all of its most important brain computers locked into a metal 'vault' of titanium, designed to take the hits instead of the computers inside. Even this 1 cm-thick layer of titanium isn't expected to reduce the radiation level down to zero; just down by about a factor of 800, which should let Juno make it through its extended mission to 2025 before too many bits are flipped and one of those bits turns out to be important.

Even with this vault equipped, Juno tends to spend very little time near Jupiter's radiation belts; the spacecraft has long orbits that take it far from the gas giant for the majority of the time it spends on its repeated journeys around Jupiter. The three principles of radiation avoidance – time, distance, and shielding – work in its favor here. But even with all three, Juno's functioning will eventually degrade, with the onboard computers gradually having their instructions scrambled bit by bit.

The problem with spacecraft is that every time you want to go somewhere different, you have to prepare for entirely different problems. All spacecraft generally have to survive the coldness of space, and some baseline level of radiation, but each target destination has its own problems for a robotic explorer.

---

* These are the Van Allen Belts, formally speaking.

The other end of the spectrum of problems is Venus. Landing on Venus doesn't come with particles slamming into your computer, it comes with a much more immediate problem.

Let's sum up spacecraft going to Venus:

Venera 4 (1967):   Crushed/melted/dead battery in the atmosphere after 100 minutes

Venera 5 (1969):   Crushed/melted in the atmosphere after 53 minutes

Venera 6 (1969):   Crushed/melted in the atmosphere after 45 minutes

Venera 7 (1970):   Crushed/melted on the surface 20 minutes after landing

Venera 8 (1972):   Crushed/melted on the surface 50 minutes after landing

Venera 9 (1975):   Crushed/melted on the surface 53 minutes after landing

Venera 10 (1975):   Crushed/melted on the surface 65 minutes after landing

Venera 11 (1978):   Crushed/melted on the surface 95+ minutes after landing

Venera 12 (1978):   Crushed/melted on the surface 110+ minutes after landing

Venera 13 (1981):   Crushed/melted on the surface 127 minutes after landing

Venera 14 (1981): Crushed/melted on the surface 57 minutes
after landing

Vega 1 (1985): Crushed/melted on the surface 54 minutes
after landing

Vega 2 (1985): Crushed/melted on the surface 54 minutes
after landing

You might notice some patterns. It turns out it's hard to make electronics work for longer than an hour when it's nearly 100 times the pressure of Earth's atmosphere and also 450 C outside. None of these outcomes could have been prevented by a big metal vault to protect your computers. All of the Venera and Vega missions were run by the former Soviet Union, and overall they were fantastically successful. These thirteen missions confirmed that Venus has no water, that it would 100% kill a human and will 100% kill a robot, and they gave us the only pictures we have of the surface of Venus.

You might also notice that the most recent date in the list is 1985, and that's when we stopped sending robots to go melt on Venus's surface. This is a real shame, because in spite of their inevitable demise, we were learning neat stuff about Venus's atmosphere from the descent phase of these robots, and more neat stuff from their records of sitting on the surface slowly melting into slag.* The European Space Agency just has to promise if they ever send a lander not to make an adorable

---

* There's no real concern for how to dispose of your spacecraft on the surface of Venus. Just let it melt.

cartoon about it, because while Philae could 'go to sleep' it's a lot harder to make a fairytale ending for a spacecraft out of 'and then it melted'.

The next lander* to Venus that's been proposed by any space agency isn't meant to launch until 2029, and the optimistic hope would be that it could last for three hours. The challenges of exploring the solar system are many and varied, and the fact that we've had so many missions go so well is a testament to the creativity of our space-exploring robot-builders.

---

* There are orbiters planned which would launch sooner than this.

EUROPA MIGHT GLOW
IN THE DARK

There's something infinitely pleasing about glowing things. Whether it's a glow stick on a summer night, or the glow-in-the-dark stars tacked to the ceiling of a child's bedroom, or simply the mesmerizing flicker of embers in a nighttime fire, humans like things that glow.

For us, they're most commonly toys, but glow-in-the-darkness shows up a lot in the natural world, without any human innovation required. I am going to immediately excuse the Sun from this conversation, because while it is certainly an exceptional example of glowing in a dark place, it's kind of cheating to use the Sun as a glow-in-the-dark object, since it's what makes us have the not-dark* in the first place.

The Moon can also be excused, because while it's a lot *closer* to glowing in the dark, in that we can see it brightly at night, it's not really *glowing*. It's a terrible mirror in the sky, is what the Moon is. It's far enough away from the Earth that it can simply catch light from the Sun that missed the daytime side of the Earth, and bounce a small fraction of that light back towards us. ('Small' being the operative word – a good mirror can reflect more than 95% of the light that hits it; the Moon,

---

* I guess if you want to be technical about it, this would be 'day'.

on the other hand, bounces back a flimsy 12% of the sunlight that hits its surface.) Inefficient spherical mirrors in the sky aren't *glowing*.

What we want to have, to call something glowing, is some object that, without an obvious source of illumination, creates its own light. Ideally, for us to notice it, it should create enough light in the visible range so that it can be detected by the human eyeball. Creating light in some other wavelength of light is all well and good, but not nearly as delightful as something you can *see*. It's also generally less dangerous; something that produces a lot of ultraviolet light would be 1) largely invisible and 2) really bad for your ability not to get a sunburn. On the other hand, things that produce a lot of infrared light would be 1) largely invisible and 2) really bad for your ability not to get a regular old burn.

So what we want, in searching for glow-in-the-dark objects in the natural world, are things that are luminescent – they produce light, some way or another.

Things that glow under a black light (a UV light, which produces a bit of blue/purple light alongside mostly the UV), but which stop glowing as soon as you switch the light off, are considered fluorescent. The object is taking in the UV light, and spitting it back out as visible light, but as soon as you run out of UV light, the visible light stops too. This category includes jellyfish, emeralds, rubies, and – apparently – all scorpions. (Scorpion fluorescence is convenient if you are in the desert and would like to play the occasionally high-stakes game of 'How far away is the nearest venomous arachnid?') If the glow persisted after turning off the light, then you'd be dealing with phosphorescence instead, which is where our glow-in-the-dark stars fall.

We also have bioluminescence as an option, where light is produced within a living creature. Fireflies are a perfect example of this one, but they're joined by anglerfish, and some marine plankton. A particular class of plankton – dinoflagellates – is responsible for a blue glow in some tropical waters. They can also cause a red tide with no glow and much fish murder, so they're not universally grand to have around.

Aside from those three options, a glow stick itself is a reaction between two chemicals, one held within a glass tube that breaks when you bend the stick, and one outside of that glass tube. This chemiluminescence will last for as long as the chemical reaction does. And then we have the way that we used to create glow-in-the-dark watch faces, back in the day, which is radioluminescence. Here, the 'radio' is the radio in radioactive, and you may well imagine (and you would be correct) that this was not a great thing to be employed to perform. The radioactive substance falls apart, and the pieces of the atom bombard the contents of the rest of the paint, and produce a glow. There are plenty more ways to make a glow on Earth, but these are the most common ones.

And none of them are able to explain how you get a glow-in-the-dark moon of Jupiter. Jupiter may instead be playing a different game entirely with the ice-encrusted moon Europa. Like phosphorescence, the moon could be glowing long after the sunlight disappears, but the *why* would be very different.

Let's be clear that there's a very strong *maybe* attached to Europa glowing, because at the moment, these are predictions from a laboratory on Earth. Many interesting scientific findings start out this way, because laboratories can often give us a sense of what to look for, before we go to a new location in our solar system. We have already looked for a glow from Europa, and not found anything, but it's not clear that's the end of the story, because we're looking from Earth, and Europa is quite far away from us.

But the mechanism would be this one: Jupiter has a tremendous magnetic field, and one of the features of a tremendous magnetic field is that you can make very tiny charged particles go very, very fast, and when they strike something (say, for example, Europa), there's a lot of energy released in the stopping, and some of that energy can become visible light, depending on what material it hits. This is the same fundamental mechanism that was used in the old, cube-shaped cathode-ray tube televisions:* speed up an electron, and then fling that electron at the screen, which is covered in materials which glow in different colors.

It's been known since the 1950s that if you strike ice with radiation like this, it will glow, so Europa is a good target for potential glow-in-the-dark-moon status, as it is *profoundly* covered in ice.

Europa's surface isn't likely to be pure water ice; most likely it has some salts mixed in, though how many and what type are still to be determined. So a group of scientists decided to place ice, mixed with the salts that Europa could have, in

---

* They're so valueless these days that sometimes you have to pay someone to remove them as garbage.

a cold chamber* and then fire electrons at it, to see what happened.

A glow happened. And in addition to being brighter the more electrons were fired, the glow changed colors depending on the type of salts that were mixed in, making Europa, potentially, a salt- and radiation-sensitive glow-in-the-dark mood ring.

Regular old water ice gave them a green glow; but adding one part Epsom salts (magnesium sulfate) to seven parts water ice created a whiter, and brighter, glow. Adding table salt to the ice made the glow fainter, and produced a more complex set of colors.† This experiment gives a new, specific prediction for what Europa's surface might be doing, if we were to go and look. And indeed going and looking is on the cards! The Europa Clipper is an upcoming NASA mission to visit this moon, hopefully in 2026, and it should be carrying a visible light camera, so we can test this prediction directly before too long.

If we do see a glow, then we can refine the Earth-based experiments to figure out exactly what mixture of ices may

. . . .

* The equipment they built to do this they named the 'Ice Chamber for Europa's High-Energy Electron and Radiation Environment Testing', which is a victim of astronomers' desire for catchy acronyms, in this case: ICE-HEART.

† The peak wavelength is a frankly upsettingly neon green color with the hexadecimal color code of #4AFF00, if you want to check yourself. Epsomite has a color peaking at #C3FF00, and table salt at #70FF00. These aren't the colors as the eye would perceive them, which is an average of all the colors of visible light the glowing ice produces, but rather the most frequently produced color of light, among that assortment.

be present; if we don't see a glow then we've learned something else about how a water ice mixture behaves near Jupiter.
Perhaps we'll see nothing. But perhaps Europa Clipper will
capture a faint shimmer from an icy moon – an irradiated, cold
world, but with a fairy-like blue-green glow in the darkness
of space.

SATURN'S RINGS ARE
FALLING APART

If the solar system had a tourism board, once they were done gloating about Earth's delightfully varied life forms, they'd probably turn to the fantastic vistas provided by Saturn's rings.

A quick summary of Saturn's main features vs. the Earth, courtesy of NASA:

	Saturn	Earth
Number of natural satellites	82	1
Planetary ring system	Yes	No

Nothing is quite as iconic in our solar system as the rings of Saturn. Visible even through a small telescope, if you want to make an instantly recognizable cartoon of Saturn, all you have to do is draw a circle with big rings around it.

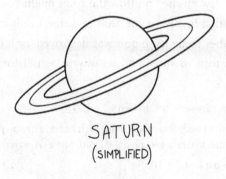

SATURN
(SIMPLIFIED)

However, if you'd like to know more about the rings of Saturn, you start running into big question marks pretty rapidly.

Things we do know:
- The rings are almost entirely water ice.*
- The rings are exceptionally thin. At their thickest, they're about a kilometer tall, but more typically only about 10 meters.
- The rings are made of trillions of small pieces of ice, from boulder-sized frozen lumps to microscopically fine ice dust.
- Their total current mass is about 40% the mass of Mimas,† one of Saturn's 82 moons.‡

Thing we do not know:
- How they got there.
- How old they are.

Thing we do know, part II:
- They're not going to stay there.

We have a few guesses on how the rings might have formed, most of which involve some kind of Large Event in Saturn's history. Either some ice moon was destroyed, or a mostly-ice moon was torn to shreds on its way to a collision with the

....

* Somewhere between 95% and 99%.

† You know, the Death Star moon. Mimas. (It has an extremely large crater on one side that makes it resemble the Death Star's laser arrangement.)

‡ This works out to 1.5×10^{19} kg.

parent planet, or Saturn shredded a big comet, or the planet was just born that way, and coalesced out of the forming solar system with a tremendously massive ring system. The problem with any kind of estimate on how long they've been there, is that these different models also come with their own guess as to how old the rings are, how massive they were when they formed, and how much of the rings had been lost before we humans sent a robot around Saturn to go measuring things directly.

Pretty much every spectacular image you've seen of Saturn comes from one very particular robot: the Cassini spacecraft. Cassini orbited Saturn for thirteen years, and when it had reached the end of its mission, it was due to plunge into the planet's atmosphere, burning itself up on our instructions. However, just before doing that, the mission scientists decided to take a few *calculated risks* with their beloved spacecraft. In particular, they wanted to send it flying through the gap between Saturn's rings and the surface of the cloud layer, a place considered too hazardous for earlier in the mission, but for a spacecraft they were about to shake into pieces and then vaporize in Saturn's atmosphere, it was worth the risk to see if they could do it. In fact, they could – and they did, 22 times, both to see what was in this space between cloud tops and ring edge (if anything), and to get really close-up views of the tops of Saturn's clouds.

So that's good: we've got a robot that we can send careening through this tiny space between the rings and Saturn's nebulous cloud tops. How does this tell us about the unravelling of Saturn's rings?

There are two ways to take apart a ring system. One is to unravel it from its innermost edge, and have the particles fall from the rings directly down onto the planet. This requires some jostling in the ring system, so that particles lose the energy they need to stay aloft, and go gravity-diving down towards the planet. This is a very direct 'eat local' campaign that Saturn's atmosphere is committing to, and this is the piece that Cassini could measure most directly by repeatedly tearing through the 2,000 km gap* between the rings and the equator of Saturn.

But we already knew about another way to unravel the rings. The Voyager probes, which passed by Saturn in the early 1980s, provided the first evidence for what became called 'ring rain'. The smallest particles in the rings – microscopic pieces of ice – can fall onto Saturn by drifting along Saturn's magnetic field. Ultraviolet light from the Sun (aka, the bit which causes sunburns) can also cause the smallest chunks of stuff to become charged, and when you have a charge, magnetism becomes important, and so these tiny motes of dust and ice can bundle themselves along the magnetic field until they hit the surface of the planet, vaporizing into gas before they arrive. To the tune of 432 to 2,870 kilograms of water every second.

Even just with this one way of pulling apart the ring system, the rings weren't expected to last another billion years – only another 300 million years, in fact. But when Cassini went in between the rings and Saturn, it found *even more rain*.†

* Cassini had a pretty narrow needle to thread here; from the southernmost point of England to the northernmost point in Scotland is 1,000 km.

† Still defined poorly as 'chunks of mostly water falling down'.

A lot of it.

It only takes four hours for this rain to fall to the surface of the cloud layer. And if there's more rain, there's an even shorter length of time before the rings kick the visibility bucket. In fact, the combination of the 'rain' discovered tumbling out of the rings directly onto Saturn's equator, and new measurements of the 'ring rain', worked out to be about 10,000 kg of material every *second*. If you drop 10,000 kg every second out of Saturn's rings, the rings should be gone in another 100 million years.

In addition to the water, Cassini could measure the rate for some of the dust that's mixed in with the rings directly, and came up with about 5.5 kg/second, but given that we know a bit about the sizes of dust in the general area, you can extrapolate your way to a much larger number.

The dust falling into Saturn's equator, measured by Cassini, is about as fine as the soot in smoke – nanometers wide. (Just for context, 5.5 kg of 1 nanometer-sized smoke particles with the density of ice would fill a cube 21 cm* to a side; that cube would hold about one septillion tiny particles (1.4×10^{24} of them), assuming random close packing.)

Meanwhile, another of Cassini's instruments, tuned to look at what materials were falling inwards, found an abundance of methane, among other molecules. Methane is too heavy to float upwards in Saturn's atmosphere, so it must be coming down from above – from the rings.

....

* The short side of A4 letter paper.

This second instrument indicated that ices are indeed falling to the clouds of Saturn, and that this is substantial – somewhere in the range of 4,800 to 45,000 kilograms of ice falling out of the rings every second. (Adding together the 5.5 kg of dust every second, plus a mid-level guesstimate from this ice precipitation, plus a guesstimate from the ring rain of 400 to 2,800 kilograms every second, is what gets you to that 10,000 kg/s number.)

The innermost ring can't support this kind of rain for very long. At the highest possible rate, the ring could only provide for this kind of loss for another 7,000 years. At the slower end, it might have 66,000 years. In either case, this is astrophysically untenable. Most probably the innermost ring is draining material somehow from its neighboring ring.

The rings of Saturn are named A through D, with A the furthest from Saturn and reasonably easy to spot, and D the innermost ring, which is less populated with ice, and thus fainter and harder to see. (The A and B rings are the two most readily visible with a small telescope.)

The C ring is much more massive than the innermost D ring, but even the C ring can't hold up to this kind of rain for ever: the estimated time to totally drain the C ring and the D ring is somewhere between 700,000 years and 7 million years from now. If the dim C ring was once as bright as the B ring, and we know how quickly the C ring is falling apart, then we can't have had this ring system around for more than 100 million years. (One paper suggests they might be as young as 10 million years old.)

No matter how you think the rings formed, all parties agree

that they were more massive in the past. If they did indeed form 100 million years ago, life on our planet was then in the middle of the Cretaceous period. T. rex, Triceratops, and the diminutive Velociraptor were swarming on the surface of the Earth, oblivious* to the drama happening in Saturn's skies.

* After all, there weren't any dinosaur telescopes.

CERES ONCE HAD VOLCANOES THAT ERUPTED WITH SALT WATER

When NASA's Dawn spacecraft arrived at Ceres, one of the larger worlds in the asteroid belt between Mars and Jupiter, the first images it sent back were positively strange. A dark world, round and pockmarked with craters, was expected. What it turned out to be was a dark world, round and pockmarked with craters, but with points of brilliant brightness dotted around the surface.

Everyone simultaneously: 'What?'*

Carefully termed 'bright spots' by the planetary scientists working on the mission, it was quickly determined that these were not 'lights', in the sense that they weren't actually glowing – they were simply very, very reflective. As Ceres turned, the bright spots dimmed as they turned into Ceres' nighttime. A true light would have kept glowing.

Once you know that something is reflective, and also in space, the suspicions go to some kind of ice or salt. Both salts and ice are very reflective surfaces, but then you have to come up with a reason for them to be there. Ceres is close enough to the Sun that we'd expect any water ice on the surface to have evaporated away, and with no atmosphere, once that

* This included the scientists, who made comments like: 'This is truly unexpected and still a mystery to us' (lead investigator Andreas Nathues).

evaporation happens we'd expect at least some of that gas to simply depart.

Further investigation found that salt, in combination with a little water, was the most likely culprit for these bright spots. Initially, a form of magnesium sulfate (Epsom salts are a similar substance), called hexahydrate magnesium sulfate, was thought most likely to be hanging out at the surface, but later research changed the salt to sodium carbonate. Well, that's one question answered, but now we have a new one: How'd it get there?

You have functionally two possibilities with a rock in space: either the salt came in from elsewhere, or it came up from underneath the surface. If it comes in from elsewhere, it's going to have to survive the arrival process, which is, on average, not particularly gentle. In fact, it's so un-gentle that we can rule this out as an option; if we're going to have salt water arrive to Ceres, the heat generated by any kind of external salt water lump hitting the surface should evaporate the water, and wouldn't leave behind these bright spots to reflect back out into space.

Which means that the salt water must be coming up from inside of Ceres – already a fascinating conclusion, because it means that somewhere inside Ceres is both salt and water. It could be quite deep down, or just below the surface, but if it's not being abruptly *delivered*, that's the only other option. If the salty water is deep enough, it may be that some external force is required anyway, just not doing any salt delivery as such. If you have a very thick crust, then even if the salt water is trying to rise, it may never have enough force to make it to the surface. But if you punch a large divot in the surface by having some other piece of rock slam into it, the effects of that crater are bigger than just the chunk you take out of the ground. In the short term, you generate quite a lot of heat, so if you have shallow ice, you ought to be able to melt it, and perhaps get a few of these salt deposits that way.

The impact crater also tends to cause deeper fractures in the rock below it. And, if there is water down below, these fractures may form a pathway that leads the water to the surface. Once the water gets there, it can then just evaporate, leaving the salt behind. We use this exact same process of letting salt water evaporate water away in order to get sea salt out of the oceans.

These fractures would be there underneath the craters on Ceres, but that's no guarantee that the brine water would need to use them. If the water were closer to the surface, then we should see bright spots anywhere on Ceres. Instead, we see that the reflective dots are mostly in younger craters – the brightest of them are in a particularly young crater* called Occator crater. So these craters might be a crucial part of the 'bright spots on Ceres' operation.

The salt we're finding – hydrated sodium chloride – also gives us a clue that this might be a relatively recent process, to go along with our relatively recent crater. This particular salt should dry out pretty quickly, which would turn it from hydrated sodium chloride to ... dehydrated sodium chloride (regular old table salt). If we're still seeing a lot of it, then this

---

* Only 78 million years old!

brightest spot on Ceres, inside Occator crater, might still be leaking salt water *now*.

On Earth, we have a particularly dramatic series of events if cracks open up in our crust: typically a volcano, and molten lava pouring out. This is the outflowing of hot magma that was too buoyant to stay down under the crust of the Earth. But volcanoes don't *have* to spill molten rock. You can also have what's called a cryovolcano or ice volcano, and they usually leak salt water instead of piping hot lava.

Cryovolcanoes work in a similar way to lava volcanoes; the difference is that they're expected only in very cold places, where exposing liquid water or brine to the elements would lead to that brine freezing over, the way that lava solidifies into rock. This frozen brine can build up into a mound, just like a shield volcano.

Ceres does indeed have a dramatic example of one of these cryovolcanoes. Informally dubbed the Lonely Mountain* but formally named Ahuna Mons, this cryovolcano is 4 km high, and with sides at 30–40 degree angles, it'd be a steep, annoying hike up a pile of chunky debris. It's only a few hundred million years old at most, which again, considering that Ceres itself formed with the rest of the solar system some 4.5 billion years ago, is pretty fresh.

It's the only cryovolcano so obviously visible on Ceres right now, but it probably had company in the past. If you leave an ice-rich pile of rubble alone for long enough, it should slump

* Yes, that is a reference to *The Hobbit*; yes, astronomers are nerds; yes, it is the only mountain on Ceres like this.

downwards. Old cryovolcanoes on Ceres would be much harder to spot – they just won't be as tall as Ahuna Mons. A hunt for these slumped remains led to 22 more probable cryovolcanoes, but it concluded that Ceres, though *technically* cryovolcanically active, is much less so than the Earth is volcanically active.

For now, the best picture of Ceres seems to be: an anomalous little round world, drifting through the asteroid belt, full of salt and strange ices, periodically springing a slow leak which deposits salt on the surface and gives this dwarf planet a daytime water haze in its cracked craters. The current model of Ceres has *at least* a reservoir of water and salt some 40 kilometers down and hundreds of kilometers across. However its internal pieces are distributed, after all these billions of years drifting in the solar system, it's still managing to pull some interesting geological tricks and actively leak salts out onto its own surface.

TRITON ORBITS BACKWARDS
AND IS DOOMED

Neptune has fourteen known moons, all of which are named after various demigods, deities, and water-related spirits. Fitting, for a planet named after the Roman god of the sea.

But by far the largest of Neptune's moons is Triton. Triton is one of the bigger moons in the solar system – it is the seventh-largest moon orbiting another planet *anywhere* in the system. Perhaps surprisingly, considering that Neptune is so much larger than the Earth, Earth's Moon is actually bigger than Triton, coming in at number five.

But Triton's claim to fame is more than just being Large. Uniquely, it orbits Neptune the wrong way around.

Almost all moons orbit in the same direction as their planet rotates. If the two formed together, this is something we'd expect. The gas cloud that all the planets formed out of would have had a direction of rotation, and that is preserved in the direction that the planets orbit the Sun. In turn, most of the planets themselves rotate on their axis in the same direction, which is the same direction that the Sun rotates around its own axis. There are exceptions in Venus, which spins backwards, and Uranus, which spins sideways, but in both of those cases we're fairly sure something dramatic happened to those planets to tip them over. When they were first forming, they probably would have rotated the same way as the rest of the planets.

So the moons of all these planets (the large ones, anyway) tend to also orbit around their planets in the same way that the planet spins. It's a small-scale version of the planets around the Sun.

Very small moons can disobey this rule, but there's an explanation for that which doesn't keep anyone awake at night, which is that sometimes an asteroid passes too near to a planet and gets caught in the planet's gravity. This is a plausible explanation for why Mars' moons are so lumpy and irregular looking, and why their orbits around Mars are so strange. It's likely they were just asteroids that were gravitationally snagged by the red planet, and now they're stuck there until they break apart.

Triton, being the seventh-largest moon in the solar system, has no such 'I'm an asteroid' excuse. How did it come to orbit Neptune the wrong way around?

Potentially in exactly the same way as Mars' doomed moons, in the end, but the reservoir that it came from wouldn't have been the asteroid belt. It is much more likely that Triton was once a Pluto-like world, in a Pluto-like orbit. (Triton is larger than Pluto, by the by.) Much like Pluto, Triton is covered in nitrogen ice, and much like Pluto, it has an active surface. Voyager 2 is the only spacecraft to have been past both Neptune and Triton, and it saw active geysers there, which also gets it onto the relatively short list of 'geologically active moons'.

It's not so easy for a planet to capture a satellite, though, especially something as large as Triton, and so for many years, planetary scientists attempted to explain how exactly Triton could have been captured. The current best theory is that Triton, before orbiting Neptune, was one half of a pair of Pluto-like worlds, and its companion might have been more massive than itself. As the two bodies came close to Neptune, they would have both felt its gravitational pull, but also a much more rapidly changing gravitational pull from their companion. In situations like this (which behave unpredictably at best), the direction that the smaller of the two interlopers is going is much more likely to be dramatically changed.

This is a tricky situation to model because, to be successful, 1) we have to reproduce Triton's weird orbit; 2) that weird orbit is changing a lot with time; 3) there are a bunch of other moons that your model can't destroy or eject; and 4) you can't break Triton either. This set of four criteria is surprisingly more difficult to work with than you might think, and so there are still a number of potential options for how Triton might have come to orbit Neptune.

The first is that Triton was captured really early in the formation of our planetary system. Triton's orbit could then slowly settle into where we see it now, and the rest of Neptune's moons, which otherwise have a habit of getting thrown into Neptune or out of orbit, could be captured later, during a period of planetary upheaval in the solar system that saw Neptune and Uranus change from orbiting much closer to the Sun outwards to where we observe them now. Triton's capture is still considered a relatively unlikely event, all told, which does help explain why Neptune got one and Uranus didn't, but not so unlikely that it shouldn't have happened at all. Perfect.

Another set of options picks up on the fact that Neptune has far fewer moons than any of the other gas giants. With only fourteen, it's much lower than Jupiter's 79, Saturn's 82, or Uranus' 27 known moons. One way to explain that is if Neptune once did have more moons, and that in a pre-Triton time, there might have been more like 30 moons around Neptune, similar to Uranus. Triton's arrival onto the scene would have caused a bit of a mess, either flinging moons of Neptune out into the deep dark of the furthest reaches of the solar system, or crashing a bunch of other moons into each other.

If the former, Triton would have flung some moons out of orbit gravitational-slingshot-style, which would allow its own orbit to sink rapidly inwards towards Neptune. This has the benefit of *not* getting rid of some of the smaller moons that orbit furthest away from Neptune, which we still see – so our model isn't allowed to break them, fling them into Neptune, or fling them into space.

If the latter, when Triton arrived, it might have promptly thrown all the moons of Neptune into such chaos that they ground themselves into grit. There's a movie trope where someone is able to move easily through a situation, but in doing so they cause a tremendous wake of chaos behind them – this model is the gravitational equivalent of an action hero driving through a series of explosions they're directly responsible for. While fun, this has the problem that if you leave large chunks of Shredded Ex-Moon hanging around, you might accidentally break Triton, which violates rule number 4; Triton has to still exist at the end of the situation.

People are still figuring out which model might be more correct. The problem of course is that none of us are time travelers, and we can't rewind time to sort out which is the right answer. However it got there, Triton certainly is there now, and on an interestingly backwards orbit; and as difficult as it is to figure out how things got to their current situation, the future is easier to predict. If we have an orbit, which we can get just from watching Triton's path around Neptune, and we have the masses of both Triton and Neptune, we can figure out what Triton's future looks like.

It's bad news for Triton. It's going to be ripped apart by Neptune. Its orbit is such that tidal forces on Triton will gradually remove energy from the moon, causing it to spiral inwards and inwards, until it's so close to Neptune that it will come apart. There's a point in space around an object which denotes how close you can get before the surface material is less well held to itself than it is to a massive object nearby.

In other words, there's a point at which Triton will be so close to Neptune that all the individual chunks of ice and rock that make up Triton feel more gravitational pull from Neptune than they do from Triton itself, and they will then depart on a voyage to Neptune. The moon breaks apart and scatters into a million pieces, all falling into Neptune.

Triton still has some time before this happens – though how much time depends a bit on how you model the orbit and what physical forces you include.

One of the first attempts to do this came up with a number of 1.5 billion years, which promptly garnered the response,

'Well that seems a bit quick'* – so either their calculation is a bit off, or we're very lucky to be able to see Triton in its current orbit at all. With a different set of assumptions, Triton has an extra 3.6 billion years left. No matter which assumptions turn out to be correct, Triton is doomed; it's just a matter of time.

* In the paper, the language is 'probably unrealistically short'.

ACKNOWLEDGMENTS

First, thanks to Duncan Heath at Icon Books, for being on board with me writing such a whimsical book.

Thanks to Drs Nicole Gugliucci and Natalie Gosnell for science-checking sections of this, and to Tom for an early proofread.

And, of course, my deepest gratitude to Pippa, Adriana, and Matthew, for keeping me grounded while writing this in the middle of a global pandemic.

I am grateful for the Oberlin College Inter-Library Loan system, which gave me access to original publications not available elsewhere.

For all distance calculations, I use: $H_0 = 0.7$, $\Omega_m = 0.3$, $\Omega_\Lambda = 0.7$, which made use of Ned Wright's cosmology calculator: http://www.astro.ucla.edu/~wright/CosmoCalc.html

NOTES

THE UNIVERSE IS THE DIMMEST IT'S BEEN IN BILLIONS OF YEARS

the time following cosmic noon can be (and sometimes has been) called 'cosmic afternoon': https://phys.org/news/2021-02-cosmic-noon-puffy-galaxies-stars.html

in which this accounting was first attempted, in 1996: Madau P., Ferguson H.C., Dickinson M.E., Giavalisco M., Steidel C.C., Fruchter A., 1996, MNRAS, 283, 1388. doi:10.1093/mnras/283.4.1388

THE UNIVERSE IS BEIGE, ON AVERAGE

The Universe is beige, on average: Baldry I.K., Glazebrook K., Baugh C.M., Bland-Hawthorn J., Bridges T., Cannon R., Cole S., et al., 2002, ApJ, 569, 582. doi:10.1086/339477

It's white: https://www.universetoday.com/92523/what-color-is-the-milky-way-white-as-snow-not-milk/; https://www.space.com/14248-milkyway-galaxy-white-color.html

the average color should have changed over the last 13.8 billion years: see the previous section, and: https://web.archive.org/web/20060127005743/http://www.eso.org/outreach/press-rel/pr-2003/pr-34-03.html

the hexadecimal color code: https://htmlcolorcodes.com/

a footnote on the third-to-last page of their paper: Baldry I.K., Glazebrook K., Baugh C.M., Bland-Hawthorn J., Bridges T., Cannon R., Cole S., et al., 2002, ApJ, 569, 582. doi:10.1086/339477 anything '[a]s long as it's not beige': https://www.wired.com/2002/03/universe-beige-not-turquoise/

even though it didn't have the largest number of votes: https://web.archive.org/web/20051220101517/http://www.pha.jhu.edu/~kgb/cosspec/topten.htm

THE GALAXY IS FLATTER THAN A CREDIT CARD

the scale *length* of the galaxy is about 10,000 light years: https://astronomy.swin.edu.au/cosmos/T/thin+disk

the scale height of the thick disk is 1,000 light years: https://astronomy.swin.edu.au/cosmos/T/thick+disk

'Petrosian' radius, which is roughly an 80% contour: https://www.sdss.org/dr12/algorithms/magnitudes/#mag_petro

the thickness of letter paper, while it varies, is typically about 0.1 mm: https://hypertextbook.com/facts/2001/JuliaSherlis.shtml

the galaxy could be as much as 100,000 light years from center to edge: López-Corredoira M., Allende Prieto C., Garzón F., Wang H., Liu C., Deng L., 2018, A&A, 612, L8. doi:10.1051/0004-6361/201832880

then we're dealing with about 1,000 light years: Rix H.-W., Bovy J., 2013, A&ARv, 21, 61. doi:10.1007/s00159-013-0061-8

there is a restaurant in New York City: https://www.thrillist.com/eat/new-york/fornino-brooklyn-nyc-biggest-pizza

held by a restaurant in Texas which is 8 feet long and 2 feet wide: https://www.guinnessworldrecords.com/world-records/largest-pizza-commercially-available

GALAXY COLLISIONS DON'T ACTUALLY CAUSE ANY STARS TO COLLIDE

0.04 solar masses-worth of stars per cubic parsec: Bovy J., 2017, MNRAS, 470, 1360. doi:10.1093/mnras/stx1277

counting stars by how massive they are (which can be done): http://www.pas.rochester.edu/~emamajek/memo_star_dens.html

Andromeda, currently coming at us at 300 kilometers per second: Peñarrubia J., Ma Y.-Z., Walker M.G., McConnachie A., 2014, MNRAS, 443, 2204. doi:10.1093/mnras/stu879

Andromeda has about a trillion stars: https://www.spitzer.caltech.edu/image/ssc2006-14a2-stars-in-andromeda

some estimates put it only a factor of two different: Peñarrubia J., Ma Y.-Z., Walker M.G., McConnachie A., 2014, MNRAS, 443, 2204. doi:10.1093/mnras/stu879

The Sun is 695,508 kilometers in radius: https://solarsystem.nasa.gov/solar-system/sun/by-the-numbers/

the densities of stars can be extremely high; 30 million per cubic light year: http://www.astronomy.ohio-state.edu/~ryden/ast162_7/notes31.html

THE GALACTIC CENTER TASTES OF RASPBERRIES AND SMELLS OF RUM

The galactic center tastes of raspberries and smells of rum: Belloche A., Garrod R.T., Müller H.S.P., Menten K.M., Comito C., Schilke P., 2009, A&A, 499, 215. doi:10.1051/0004-6361/200811550

The density of air can be described in terms of kilograms per cubic meter: https://www.grc.nasa.gov/www/k-12/BGP/airprop.html

we can use Loschmidt's constant: https://en.wikipedia.org/wiki/Loschmidt _constant

a heaping *one* atom per cubic centimeter: http://www.outerspacecentral. com/ism_page.html

at around about 100 atoms per cubic centimeter: Goldsmith P.F., Lis D.C., Hills R., Lasenby J., 1990, ApJ, 350, 186. doi:10.1086/168372

Ethyl alcohol has been found here too: Zuckerman B., Turner B.E., Johnson D.R., Clark F.O., Lovas F.J., Fourikis N., Palmer P., et al., 1975, ApJL, 196, L99. doi:10.1086/181753

Formaldehyde (causes cancer): https://www.cancer.org/cancer/cancer-causes/formaldehyde.html

formic acid (will burn you): https://pubchem.ncbi.nlm.nih.gov/compound/ Formic-acid

n-propyl cyanide (forms regular old toxic cyanide in the body): https:// www.cdc.gov/niosh/npg/npgd0086.html

and methanol: Zuckerman B., Turner B.E., Johnson D.R., Clark F.O., Lovas F.J., Fourikis N., Palmer P., et al., 1975, ApJL, 196, L99. doi:10.1086/181753

toxic if you breathe, touch, or swallow it, and also catches fire: https:// pubchem.ncbi.nlm.nih.gov/compound/Methanol#section=Safety-and -Hazards

to name a few: https://analyticalscience.wiley.com/do/10.1002/sepspec. 21408ezine

THE CENTERS OF GALAXIES CAN BLOW GALAXY-SIZED BUBBLES

if the galaxy is massive (like the Milky Way): McNamara B.R., Nulsen P.E.J., 2007, ARA&A, 45, 117. doi:10.1146/annurev.astro.45.051806.110625

In 2010, we found that our own Milky Way has two massive bubbles: Su M., Slatyer T.R., Finkbeiner D.P., 2010, ApJ, 724, 1044. doi:10.1088/0004-637X/724/2/1044

they're about 50,000 light years in length: https://fermi.gsfc.nasa.gov/science/constellations/pages/bubbles.html

the bubbles were clearly visible: Su M., Slatyer T.R., Finkbeiner D.P., 2010, ApJ, 724, 1044. doi:10.1088/0004-637X/724/2/1044

'sometime in the last ten million years': Su M., Slatyer T.R., Finkbeiner D.P., 2010, ApJ, 724, 1044. doi:10.1088/0004-637X/724/2/1044

to generate the energy required in 1,000–10,000 years: Su M., Slatyer T.R., Finkbeiner D.P., 2010, ApJ, 724, 1044. doi:10.1088/0004-637X/724/2/1044

In this model: Mou G., Yuan F., Bu D., Sun M., Su M., 2014, ApJ, 790, 109. doi:10.1088/0004-637X/790/2/109

one model suggests that they may only be a million years old: https://arxiv.org/pdf/1207.4185.pdf

something dramatic some 3.5 million years ago: https://iopscience.iop.org/article/10.3847/1538-4357/ab44c8/meta#apjab44c8s5

at least 50,000 parsecs away: Fox A.J., Frazer E.M., Bland-Hawthorn J., Wakker B.P., Barger K.A., Richter P., 2020, ApJ, 897, 23. doi:10.3847/1538-4357/ab92a3

One particular model: Fox A.J., Frazer E.M., Bland-Hawthorn J., Wakker B.P., Barger K.A., Richter P., 2020, ApJ, 897, 23. doi:10.3847/1538-4357/ab92a3

'mysterious physical origin': Yang H.-Y., Ruszkowski M., Zweibel E., 2018, Galax, 6, 29. doi:10.3390/galaxies6010029

A DISTANT BLACK HOLE IS SURROUNDED BY WATER

A distant black hole is surrounded by water: Bradford C.M., Bolatto A.D., Maloney P.R., Aguirre J.E., Bock J.J., Glenn J., Kamenetzky J., et al., 2011, ApJL, 741, L37. doi:10.1088/2041-8205/741/2/L37; Lis D.C., Neufeld D.A., Phillips T.G., Gerin M., Neri R., 2011, ApJL, 738, L6. doi:10.1088/2041-8205/738/1/L6

First detected in a radio survey as an unresolved, bright spot: Matthews T.A., Sandage A.R., 1963, ApJ, 138, 30. doi:10.1086/147615; Schmidt M., 1963, Nature, 197, 1040. doi:10.1038/1971040a0

the oldest glowing water we've yet found: Lis D.C., Neufeld D.A., Phillips T.G., Gerin M., Neri R., 2011, ApJL, 738, L6. doi:10.1088/2041-8205/738/1/L6

it was the brightest object in the known universe: Irwin M.J., Ibata R.A., Lewis G.F., Totten E.J., 1998, ApJ, 505, 529. doi:10.1086/306213

25,000 times the mass of the Sun: Bradford C.M., Bolatto A.D., Maloney P.R., Aguirre J.E., Bock J.J., Glenn J., Kamenetzky J., et al., 2011, ApJL, 741, L37. doi:10.1088/2041-8205/741/2/L37

140 trillion times all the water in the world's ocean: https://www.nasa.gov/topics/universe/features/universe20110722.html

the largest so far is 'only' about 200 times the mass of the Sun: Bestenlehner J.M., Crowther P.A., Caballero-Nieves S.M., Schneider F.R.N., Simón-Díaz S., Brands S.A., de Koter A., et al., 2020, MNRAS, 499, 1918. doi:10.1093/mnras/staa2801

it's still a pretty frozen -53 Celsius: https://www.nasa.gov/topics/universe/features/universe20110722.html

a low-frequency one which gets called a *maser*: https://astronomy.swin.edu.au/cosmos/m/Masers

SOME GALAXIES LOOK LIKE JELLYFISH

they wound up in their own box labeled 'irregular' or 'peculiar': Arp H., 1966, ApJS, 14, 1. doi:10.1086/190147

tens to hundreds of thousands of light years away from the disk: Poggianti B.M., Jaffé Y.L., Moretti A., Gullieuszik M., Radovich M., Tonnesen S., Fritz J., et al., 2017, Nature, 548, 304. doi:10.1038/nature23462

ram pressure stripping: Gunn J.E., Gott J.R., 1972, ApJ, 176, 1. doi:10.1086/151605

inside those rapidly removed streamers, we can form new stars: Rawle T.D., Altieri B., Egami E., Pérez-González P.G., Richard J., Santos J.S., Valtchanov I., et al., 2014, MNRAS, 442, 196. doi:10.1093/mnras/stu868

the Very Large Array in New Mexico: Owen F.N., Keel W.C., Wang Q.D., Ledlow M.J., Morrison G.E., 2006, AJ, 131, 1974. doi:10.1086/500573

it's much easier to see them: https://esahubble.org/images/heic1404b/

The first identified jellyfish galaxies: Ebeling H., Stephenson L.N., Edge A.C., 2014, ApJL, 781, L40. doi:10.1088/2041-8205/781/2/L40

Follow-up searches: Fumagalli M., Fossati M., Hau G.K.T., Gavazzi G., Bower R., Sun M., Boselli A., 2014, MNRAS, 445, 4335. doi:10.1093/mnras/stu2092; Poggianti B.M., Moretti A., Gullieuszik M., Fritz J., Jaffé Y., Bettoni D., Fasano G., et al., 2017, ApJ, 844, 48. doi:10.3847/1538-4357/aa78ed; Roberts I.D., van Weeren R.J., McGee S.L., Botteon A., Drabent A., Ignesti A., Rottgering H.J.A., et al., 2021, A&A, 650, A111. doi:10.1051/0004-6361/202140784

THE WHOLE SKY GLOWS IN NEUTRAL HYDROGEN

The whole sky glows in neutral hydrogen: Wisotzki L., Bacon R., Brinchmann J., Cantalupo S., Richter P., Schaye J., Schmidt K.B., et al., 2018, Nature, 562, 229. doi:10.1038/s41586-018-0564-6

One tiny packet of light every 10 million years: https://astronomy.swin. edu.au/cosmos/s/Spin-flip+Transition

enough of hydrogen's shimmering faint glow that it fills the entire sky: HI4PI Collaboration, Ben Bekhti N., Flöer L., Keller R., Kerp J., Lenz D., Winkel B., et al., 2016, A&A, 594, A116. doi:10.1051/0004-6361/201629178

a typical human hair might be around 60,000 nanometers in diameter: https://www.science.org.au/curious/technology-future/how-small-nanoscale -small

this Lyman alpha glow of illuminated hydrogen becomes visible: https://academo.org/demos/wavelength-to-colour-relationship/

SOME OF THE STARS IN THE GALAXY ARE JUST PASSING THROUGH

Some of the stars in the galaxy are just passing through: Marchetti T., Rossi E.M., Brown A.G.A., 2019, MNRAS, 490, 157. doi:10.1093/mnras/sty2592

The Gaia Space Telescope: https://sci.esa.int/web/gaia

this is only about 1% of the stars in the galaxy: https://sci.esa.int/web/gaia

SUPERMASSIVE BLACK HOLES CAN SING A SUPER-LOW B FLAT

Supermassive black holes can sing a super-low B flat: Fabian A.C., Sanders J.S., Allen S.W., Crawford C.S., Iwasawa K., Johnstone R.M., Schmidt R.W., et al., 2003, MNRAS, 344, L43. doi:10.1046/j.1365-8711.2003.06902.x; Fabian A.C., Sanders J.S., Taylor G.B., Allen S.W., Crawford C.S., Johnstone R.M., Iwasawa K., 2006, MNRAS, 366, 417. doi:10.1111/j.1365-2966. 2005.09896.x

The galaxy itself, called NGC 1275: http://ned.ipac.caltech.edu/cgi-bin/ objsearch?search_type=Obj_id&objid=58262&objname=1&img_stamp= YES&hconst=73.0&omegam=0.27&omegav=0.73&corr_z=1

host to some 2.43 × 10^{11} solar masses of material: Mathews W.G., Faltenbacher A., Brighenti F., 2006, ApJ, 638, 659. doi:10.1086/499119

Surrounded by a powerful magnetic field: https://esahubble.org/images/ heic0817a/

unusually shaped bubbles as a result: Hitomi Collaboration, Aharonian F., Akamatsu H., Akimoto F., Allen S.W., Angelini L., Audard M., et al., 2018, PASJ, 70, 11. doi:10.1093/pasj/psy004

240 million light years away from Earth: http://ned.ipac.caltech.edu/cgi-bin/ objsearch?search_type=Obj_id&objid=132583&objname=1&img_stamp =YES&hconst=70.0&omegam=0.30&omegav=0.70&corr_z=1

in this particular cluster, we know that the temperature isn't changing: Fabian A.C., Sanders J.S., Allen S.W., Crawford C.S., Iwasawa K., Johnstone R.M., Schmidt R.W., et al., 2003, MNRAS, 344, L43. doi:10.1046/j.1365-8711.2003.06902.x

Or a sound wave: Fabian A.C., Sanders J.S., Taylor G.B., Allen S.W., Crawford C.S., Johnstone R.M., Iwasawa K., 2006, MNRAS, 366, 417. doi:10.1111/j.1365-2966.2005.09896.x

but teenagers certainly can, and find incredibly irritating: https://www. theguardian.com/uk/2006/oct/06/science.highereducation

has a frequency of 17.5 kilohertz: https://en.wikipedia.org/wiki/The_ Mosquito

humans tend to be able to hear noises down to about 20 hertz: https:// www.ncbi.nlm.nih.gov/books/NBK10924/

creates cycles one time every 9.6 *million years*: Fabian A.C., Sanders J.S., Allen S.W., Crawford C.S., Iwasawa K., Johnstone R.M., Schmidt R.W., et al., 2003, MNRAS, 344, L43. doi:10.1046/j.1365-8711.2003.06902.x

This is, weirdly, technically a note: http://news.bbc.co.uk/2/hi/science/nature/3096776.stm

It's 57 octaves below middle C: https://www.nasa.gov/home/hqnews/2003/sep/HQ_03284_Chandra_Hears.html

is decidedly off-key: https://chandra.harvard.edu/press/06_releases/press_100506.html

SOME BLACK HOLES COULD BE NECROMANCERS

Some black holes could be necromancers: Anninos P., Fragile P.C., Olivier S.S., Hoffman R., Mishra B., Camarda K., 2018, ApJ, 865, 3. doi:10.3847/1538-4357/aadad9

somewhere between 25% and 50%: Mathieu R.D., Geller A.M., 2009, Nature, 462, 1032. doi:10.1038/nature08568; Raghavan D., McAlister H.A., Henry T.J., Latham D.W., et al., 2010, ApJS, 190, 1. doi:10.1088/0067-0049/190/1/1; Tian Z.-J., Liu X.-W., Yuan H.-B., Chen B.-Q., Xiang M.-S., Huang Y., Wang C., et al., 2018, RAA, 18, 052. doi:10.1088/1674-4527/18/5/52

have a brief stellar lifetime: http://www.ifa.hawaii.edu/users/hu/stars.html

living about 10 million years: https://www.cosmos.esa.int/web/cesar/the-hertzsprung-russell-diagram

for a period of a few weeks: http://www.ucolick.org/~woosley/ay112-14/lectures/lecture16.4x.pdf

If they are far enough away, then nothing much occurs: Liu J., Zhang H., Howard A.W., Bai Z., Lu Y., Soria R., Justham S., et al., 2019, Nature, 575, 618. doi:10.1038/s41586-019-1766-2

and leave behind a temporary (though lovely) planetary nebula: https://esahubble.org/wordbank/planetary-nebula/; https://www.nasa.gov/ mission_pages/chandra/multimedia/planetary_nebula.html; https://www. britannica.com/science/planetary-nebula

only about half as massive as the original star was: https://imagine.gsfc. nasa.gov/science/objects/dwarfs2.html

to just hang around in the Universe: https://astronomy.com/news-observing/ask%20astro/2004/12/how%20long%20does%20it%20take %20for%20a%20white%20dwarf%20to%20turn%20into%20a%20black %20dwarf

you can make your white dwarf periodically explode (a nova): https:// en.wikipedia.org/wiki/Nova

to fall straight in rather than stretching out first: Anninos P., Fragile P.C., Olivier S.S., Hoffman R., Mishra B., Camarda K., 2018, ApJ, 865, 3. doi:10.3847/1538-4357/aadad9

NEUTRON STARS COLLIDING GAVE US GOLD AND PLATINUM ON EARTH

the signature of some massive objects slamming into each other: https://www.nasa.gov/press-release/nasa-missions-catch-first-light-from-a-gravitational-wave-event

the best model that fitted the data was of two colliding neutron stars: Smartt S.J., Chen T.-W., Jerkstrand A., Coughlin M., Kankare E., Sim S.A., Fraser M., et al., 2017, Nature, 551, 75. doi:10.1038/nature24303

about 1.5 times the mass of the Sun: https://media.ligo.northwestern.edu/ gallery/mass-plot

more like 100 billion times stronger: https://astronomy.swin.edu.au/ cosmos/N/Neutron+Star

The other option would be to form a magnetar: Fong W., Laskar T., Rastinejad J., Escorial A.R., Schroeder G., Barnes J., Kilpatrick C.D., et al., 2021, ApJ, 906, 127. doi:10.3847/1538-4357/abc74a

70+ telescopes across nearly as many countries: https://www.universe today.com/137629/gw170817-update-surprises-first-gravitational-wave-observed-independently/

a few every million years or so counts: Andreoni I., Kool E.C., Sagués Carracedo A., Kasliwal M.M., Bulla M., Ahumada T., Coughlin M.W., et al., 2020, ApJ, 904, 155. doi:10.3847/1538-4357/abbf4c

accounting for somewhere around 90% of all atoms: https://education.jlab.org/itselemental/ele001.html

someone's done the math on this: Bartos I., Marka S., 2019, Nature, 569, 85. doi:10.1038/s41586-019-1113-7

SOME OBJECTS SPIN SO FAST THEY NEARLY SELF-DESTRUCT

spun on its own axis once every 1.33 seconds: Arzoumanian Z., Nice D.J., Taylor J.H., Thorsett S.E., 1994, ApJ, 422, 671. doi:10.1086/173760

Its name is B1937+21: Backer D. C., Kulkarni S.R., Heiles C., Davis M.M., Goss W.M., 1982, Nature, 300, 615. doi:10.1038/300615a0

PSR J1748–2446ad came to steal the crown: Hessels J.W.T., Ransom S.M., Stairs I.H., Freire P.C.C., Kaspi V.M., Camilo F., 2006, Sci, 311, 1901. doi:10.1126/science.1123430

an admittedly impressive 342 revolutions per minute: https://www.guinnessworldrecords.com/world-records/fastest-spin-ice-skating

moving at 13% the speed of light: Backer D.C., Kulkarni S.R., Heiles C., Davis M.M., Goss W.M., 1982, Nature, 300, 615. doi:10.1038/300615a0

must be less than 16 km from center to edge: Hessels J.W.T., Ransom S.M., Stairs I.H., Freire P.C.C., Kaspi V.M., Camilo F., 2006, Sci, 311, 1901. doi:10.1126/science.1123430

they have a companion star: Bhattacharya D., van den Heuvel E.P.J., 1991, PhR, 203, 1. doi:10.1016/0370-1573(91)90064-S

IT RAINS IRON ON SOME BROWN DWARFS

'No thank you, too hot for me' 1,000 Kelvin: https://astronomy.swin.edu.au/cosmos/B/brown+dwarf

oven or cup of coffee temperature: https://www.space.com/22659-brown-dwarf-stars-hot-as-oven.html; https://www.nationalgeographic.com/adventure/article/110323-coldest-star-discovered-cup-coffee-brown-dwarf-hawaii-space-science

below the temperature at which water would freeze: https://www.nasa.gov/press/2014/april/nasas-spitzer-and-wise-telescopes-find-close-cold-neighbor-of-sun/

the admittedly summery 3 million degrees: https://astronomy.swin.edu.au/cosmos/B/brown+dwarf

we had to wait until 1995: https://chandra.harvard.edu/xray_sources/pdf/brown_dwarfs.pdf

the thermostat is set as low as it's going to get (5,600 Celsius): https://www.nasa.gov/sites/default/files/files/Sun_Lithograph.pdf

the surface layers of two nearby brown dwarfs: Crossfield I.J.M., Biller B., Schlieder J.E., Deacon N.R., Bonnefoy M., Homeier D., Allard F., et al., 2014, Nature, 505, 654. doi:10.1038/nature12955

how the gases present in our Sun would start to behave at cooler temperatures: https://www.nytimes.com/2002/08/06/science/forecast-for-brown-dwarf-stars-iron-rain-heavy-at-times.html

become liquid as it cools at the surface: https://www.space.com/2576-wild-weather-iron-rain-failed-stars.html

Oh, and also sand: https://www.futurity.org/telescopes-reveal-brown-dwarfs-iron-rain/

Half of another sample of brown dwarfs: https://www.space.com/24192-stormy-weather-brown-dwarfs-aas223.html

'most' or 'all' of the brown dwarfs have these tremendous storms: https://www.jpl.nasa.gov/news/stormy-stars-nasas-spitzer-probes-weather-on-brown-dwarfs

bands of material rotate into and out of view: https://www.nasa.gov/feature/jpl/scientists-improve-brown-dwarf-weather-forecasts

a lot like the bands of material seen on Neptune: Apai D., Karalidi T., Marley M.S., Yang H., Flateau D., Metchev S., Cowan N.B., et al., 2017, Sci, 357, 683. doi:10.1126/science.aam9848

the clouds are more stripes and less spots: Apai D., Nardiello D., Bedin L.R., 2021, ApJ, 906, 64. doi:10.3847/1538-4357/abcb97

winds, sometimes up to 1,400 km per second: https://www.nasa.gov/feature/jpl/in-a-first-nasa-measures-wind-speed-on-a-brown-dwarf

the air cools enough and the iron could rain out: https://www.sciencedaily.com/releases/2020/03/200311121832.htm; Ehrenreich D., Lovis C., Allart R., Zapatero Osorio M.R., Pepe F., Cristiani S., Rebolo R., et al., 2020, Nature, 580, 597. doi:10.1038/s41586-020-2107-1

enough to permit iron to condense out into rain: Savel A.B., Kempton E.M.-R., Malik M., Komacek T.D., Bean J.L., May E.M., Stevenson K.B., et al., 2021, arXiv, arXiv:2109.00163; Wardenier J.P., Parmentier V., Lee E.K.H., Line M.R., Gharib-Nezhad E., 2021, MNRAS, 506, 1258. doi:10.1093/mnras/stab1797

WE SAW A CHUNK OF ROCK OR ICE FROM OUTSIDE THE SOLAR SYSTEM

The first was detected in 2017: Meech K.J., Weryk R., Micheli M., Kleyna J.T., Hainaut O.R., Jedicke R., Wainscoat R.J., et al., 2017, Nature, 552, 378. doi:10.1038/nature25020

in consultation with Ka'iu Kimura and Larry Kimura: http://www.hawaii. edu/news/2017/11/20/an-interstellar-visitor-unmasked/

'sent from the distant past to reach out to us': https://www.nature.com/ articles/nature25020

named Borisov after its discoverer: https://minorplanetcenter.net/mpec/ K19/K19S72.html

We had about two weeks to observe 'Oumuamua: Meech K.J., Weryk R., Micheli M., Kleyna J.T., Hainaut O.R., Jedicke R., Wainscoat R.J., et al., 2017, Nature, 552, 378. doi:10.1038/nature25020

It was also tumbling chaotically: Fraser W.C., Pravec P., Fitzsimmons A., Lacerda P., Bannister M.T., Snodgrass C., Smolić I., 2018, NatAs, 2, 383. doi:10.1038/s41550-018-0398-z; Drahus M., Guzik P., Waniak W., Handzlik B., Kurowski S., Xu S., 2018, NatAs, 2, 407. doi:10.1038/ s41550-018-0440-1

comets or icy asteroids within our solar system: Bannister M.T., Schwamb M.E., Fraser W.C., Marsset M., Fitzsimmons A., Benecchi S.D., Lacerda P., et al., 2017, ApJL, 851, L38. doi:10.3847/2041-8213/aaa07c

It seemed to be behaving like a rocky asteroid: Meech K.J., Weryk R., Micheli M., Kleyna J.T., Hainaut O.R., Jedicke R., Wainscoat R.J., et al., 2017, Nature, 552, 378. doi:10.1038/nature25020

more elongated than anything we know of in our solar system: Meech K.J., Weryk R., Micheli M., Kleyna J.T., Hainaut O.R., Jedicke R., Wainscoat R.J., et al., 2017, Nature, 552, 378. doi:10.1038/nature25020

faster than we expected if it were just a rock: Micheli M., Farnocchia D., Meech K.J., Buie M.W., Hainaut O.R., Prialnik D., Schörghofer N., et al., 2018, Nature, 559, 223. doi:10.1038/s41586-018-0254-4

that much gas might have been invisible to the telescopes observing earlier: Micheli M., Farnocchia D., Meech K.J., Buie M.W., Hainaut O.R., Prialnik D., Schörghofer N., et al., 2018, Nature, 559, 223. doi:10.1038/s41586-018-0254-4

it had a little bit of a protective coating: Fitzsimmons A., Snodgrass C., Rozitis B., Yang B., Hyland M., Seccull T., Bannister M.T., et al., 2018, NatAs, 2, 133. doi:10.1038/s41550-017-0361-4

flung out of another solar system while it was just forming planets: Portegies Zwart S., Torres S., Pelupessy I., Bédorf J., Cai M.X., 2018, MNRAS, 479, L17. doi:10.1093/mnrasl/sly088

that got too close to its star and was destroyed: Zhang Y., Lin D.N.C., 2020, NatAs, 4, 852. doi:10.1038/s41550-020-1065-8

remains of some object which was once like Pluto: Jackson A.P., Desch S.J., 2021, JGRE, 126, e06706. doi:10.1029/2020JE006706

articles with titles like "Oumuamua is not Artificial': Katz J.I., 2021, arXiv, arXiv:2102.07871

'no compelling evidence to favor an alien explanation for 'Oumuamua': 'Oumuamua ISSI Team, Bannister M.T., Bhandare A., Dybczyński P.A., Fitzsimmons A., Guilbert-Lepoutre A., Jedicke R., et al., 2019, NatAs, 3, 594. doi:10.1038/s41550-019-0816-x

IO HAS LAKES OF LAVA

the (many) amorous attentions of Zeus: https://www.britannica.com/topic/Io-Greek-mythology

Io itself is a touch larger than Earth's Moon: https://solarsystem.nasa.gov/
moons/jupiter-moons/io/overview/

more than 100 kilometers up away from its surface into space: https://
www.jpl.nasa.gov/images/active-volcanic-plumes-on-io

Io completes one rotation every time it completes one orbit: https://
solarsystem.nasa.gov/moons/jupiter-moons/io/overview/

hundreds and hundreds of volcanoes: Lopes R.M.C., Kamp L.W., Smythe
W.D., Mouginis-Mark P., Kargel J., Radebaugh J., Turtle E.P., et al., 2004,
Icar, 169, 140. doi:10.1016/j.icarus.2003.11.013

and some intermittently active: de Kleer K., de Pater I., Molter E.M.,
Banks E., Davies A.G., Alvarez C., Campbell R., et al., 2019, AJ, 158, 29.
doi:10.3847/1538-3881/ab2380

has been an active lava lake since 1979: Rathbun J.A., Spencer
J.R., Davies A.G., Howell R.R., Wilson L., 2002, GeoRL, 29, 1443.
doi:10.1029/2002GL014747

and hot lava from beneath sweeping across it: https://www.space.
com/36788-lava-waves-jupiter-volcanic-moon-io.html

it ought to have largely filled in the hole it lives in: Rathbun J.A., Spencer
J.R., Davies A.G., Howell R.R., Wilson L., 2002, GeoRL, 29, 1443.
doi:10.1029/2002GL014747

and 'generally smaller plumes': Lopes R.M.C., Kamp L.W., Smythe W.D.,
Mouginis-Mark P., Kargel J., Radebaugh J., Turtle E.P., et al., 2004, Icar,
169, 140. doi:10.1016/j.icarus.2003.11.013

and Tvashtar about the same: https://www.jpl.nasa.gov/images/galileos-
last-view-of-tvashtar-io; https://science.sciencemag.org/content/318/5848/
240

anywhere between 355 and 415 km high: https://photojournal.jpl.nasa.
gov/catalog/PIA02588

The tallest plume I was able to find: https://www.jpl.nasa.gov/images/northern-plume-and-plume-deposits-on-io

the International Space Station (ISS) orbits at about 400 km: https://www.nasa.gov/mission_pages/station/expeditions/expedition26/iss_altitude.html

over a 700 km radius of the surrounding region: https://www.jpl.nasa.gov/images/galileos-last-view-of-tvashtar-io

If this were Earth, and it were centered on London: https://www.mapdevelopers.com/draw-circle-tool.php?circles=%5B%5B700000%2C51.5073219%2C-0.1276474%2C%22%23AAAAAA%22%2C%22%23000000%22%2C0%5D%5D

tremendous pyroclastic explosions: https://www.nationalgeographic.org/encyclopedia/pyroclastic-flow/

sometimes described as a 'glowing avalanche': Sigurdsson, H., Cashdollar, S., and Stephen R.J. Sparks. (1982). The Eruption of Vesuvius in A.D. 79: Reconstruction from Historical and Volcanological Evidence. *American Journal of Archaeology*, 86(1), 39–51. https://doi.org/10.2307/504292

anything like the volcanoes on Earth: Lopes R.M.C., Kamp L.W., Smythe W.D., Mouginis-Mark P., Kargel J., Radebaugh J., Turtle E.P., et al., 2004, Icar, 169, 140. doi:10.1016/j.icarus.2003.11.013

IT RAINS DIAMONDS ON NEPTUNE

light takes four hours and nine minutes to reach Neptune: https://solarsystem.nasa.gov/planets/neptune/overview/

clocking in at some 1,100+ miles an hour: Kaspi Y., Showman A.P., Hubbard W.B., Aharonson O., Helled R., 2013, Nature, 497, 344. doi:10.1038/nature12131; https://svs.gsfc.nasa.gov/11349

both Voyager 2 and the Hubble Space Telescope: Hammel, H.B., et al. 'Neptune's Wind Speeds Obtained by Tracking Clouds in Voyager Images.'

Science, vol. 245, no. 4924, 1989, pp. 1367–9. *JSTOR*, www.jstor.org/ stable/1704269, accessed June 16th, 2021; Smith, B.A., et al. 'Voyager 2 at Neptune: Imaging Science Results.' *Science*, vol. 246, no. 4936, 1989, pp. 1422–49. *JSTOR*, www.jstor.org/stable/1704900, accessed June 16th, 2021.

it measured how reflective the planet was: Pearl J.C., Conrath B.J., 1991, JGR, 96, 18921. doi:10.1029/91JA01087

its distinctive blue color comes from methane: https://www.britannica. com/science/methane

freeing up the carbon to make diamonds: Ross M., 1981, Nature, 292, 435. doi:10.1038/292435a0

Propane is one: https://www.merriam-webster.com/dictionary/propane

the other one we know is polystyrene: https://www.britannica.com/science/ polystyrene

you can fire a laser at polystyrene: Kraus D., Vorberger J., Pak A., Hartley N.J., Fletcher L.B., Frydrych S., Galtier E., et al., 2017, NatAs, 1, 606. doi:10.1038/s41550-017-0219-9

it's hydrogen, helium, and methane, as we saw: https://solarsystem.nasa. gov/planets/neptune/by-the-numbers/

It's also less ... explosive than methane: https://www.britannica.com/ science/methane

In the most recent study on this: Frydrych S., Vorberger J., Hartley N.J., Schuster A.K., Ramakrishna K., Saunders A.M., van Driel T., et al., 2020, NatCo, 11, 2620. doi:10.1038/s41467-020-16426-y

plasma, which then effectively explodes outwards: https://www.american scientist.org/article/on-neptune-its-raining-diamonds

Are there diamond icebergs?: https://gizmodo.com/scientists-say-its-raining-diamonds-on-neptune-and-uran-1798150640

AN EXOPLANET WE THOUGHT WAS MADE OF DIAMOND MIGHT BE LAVA INSTEAD

if you smash silicon carbide in water between two diamonds: Allen-Sutter H., Garhart E., Leinenweber K., Prakapenka V., Greenberg E., Shim S.-H., 2020, PSJ, 1, 39. doi:10.3847/PSJ/abaa3e

you get more diamond and silica: https://www.cnet.com/news/scientist s-investigate-diamond-planets-unlike-anything-in-our-solar-system/

most excitedly proclaimed to be Diamond Planet: Madhusudhan N., Lee K.K.M., Mousis O., 2012, ApJL, 759, L40. doi:10.1088/2041-8205/759/2/L40

the fourth planet discovered around the star 55 Cancri: Winn J.N., Matthews J.M., Dawson R.I., Fabrycky D., Holman M.J., Kallinger T., Kuschnig R., et al., 2011, ApJL, 737, L18. doi:10.1088/2041-8205/737/1/L18

orbits it once every eighteen hours: Dawson R.I., Fabrycky D.C., 2010, ApJ, 722, 937. doi:10.1088/0004-637X/722/1/937

a surface temperature of 2,700 Kelvin: Demory B.-O., Gillon M., de Wit J., Madhusudhan N., Bolmont E., Heng K., Kataria T., et al., 2016, Nature, 532, 207. doi:10.1038/nature17169

Mercury's orbit around the Sun takes 88 days: https://solarsystem.nasa.gov/planets/mercury/overview/

bone-dry, with a typical surface temperature of 430 C: https://solar system.nasa.gov/planets/mercury/in-depth/

overall, it's about as dense as lead: https://www.space.com/11544-densest-alien-planet-55cancrie.html

the temporary delights of a diamond planet: Madhusudhan N., Lee K.K.M., Mousis O., 2012, ApJL, 759, L40. doi:10.1088/2041-8205/759/2/L40

'Yes, atmosphere': Angelo I., Hu R., 2017, AJ, 154, 232. doi:10.3847/1538-3881/aa9278

'Not wet': Bourrier V., Dumusque X., Dorn C., Henry G.W., Astudillo-Defru N., Rey J., Benneke B., et al., 2018, A&A, 619, A1. doi:10.1051/0004-6361/201833154

'Probably medium normal rock?': Bourrier V., Dumusque X., Dorn C., Henry G.W., Astudillo-Defru N., Rey J., Benneke B., et al., 2018, A&A, 619, A1. doi:10.1051/0004-6361/201833154

Rock is *not* an efficient system, even when molten: Angelo I., Hu R., 2017, AJ, 154, 232. doi:10.3847/1538-3881/aa9278

hottest *not* right where its star is shining directly above it: Angelo I., Hu R., 2017, AJ, 154, 232. doi:10.3847/1538-3881/aa9278

the day/night temperature switch would be less intense: Bourrier V., Dumusque X., Dorn C., Henry G.W., Astudillo-Defru N., Rey J., Benneke B., et al., 2018, A&A, 619, A1. doi:10.1051/0004-6361/201833154; Demory B.-O., Gillon M., de Wit J., Madhusudhan N., Bolmont E., Heng K., Kataria T., et al., 2016, Nature, 532, 207. doi:10.1038/nature17169

'more similar to water at room temperature than to solid rock': Demory B.-O., Gillon M., de Wit J., Madhusudhan N., Bolmont E., Heng K., Kataria T., et al., 2016, Nature, 532, 207. doi:10.1038/nature17169

'Does it rain lava?': Brandeker A., Alibert Y., Bourrier V., Delrez L., Demory B.-O., Ehrenreich D., Fridlund M., et al., 2021, jwst.prop, 2084

So, the skies would sparkle: https://exoplanets.nasa.gov/eyes-on-exoplanets/#/planet/55_Cnc_e/

THERE'S A PITCH-BLACK EXOPLANET

Zooming around its host star once every 26 hours: Hebb L., Collier-Cameron A., Loeillet B., Pollacco D., Hébrard G., Street R.A., Bouchy F., et al., 2009, ApJ, 693, 1920. doi:10.1088/0004-637X/693/2/1920

a planet as black as tar: Bell T.J., Nikolov N., Cowan N.B., Barstow J.K., Barman T.S., Crossfield I.J.M., Gibson N.P., et al., 2017, ApJL, 847, L2. doi:10.3847/2041-8213/aa876c

does seem to have two red dwarf companions: Bechter E.B., Crepp J.R., Ngo H., Knutson H.A., Batygin K., Hinkley S., Muirhead P.S., et al., 2014, ApJ, 788, 2. doi:10.1088/0004-637X/788/1/2; Bergfors C., Brandner W., Daemgen S., Biller B., Hippler S., Janson M., Kudryavtseva N., et al., 2013, MNRAS, 428, 182. doi:10.1093/mnras/sts019

At the time of its discovery in 2009: Bechter E.B., Crepp J.R., Ngo H., Knutson H.A., Batygin K., Hinkley S., Muirhead P.S., et al., 2014, ApJ, 788, 2. doi:10.1088/0004-637X/788/1/2

just a bit more massive: Hebb L., Collier-Cameron A., Loeillet B., Pollacco D., Hébrard G., Street R.A., Bouchy F., et al., 2009, ApJ, 693, 1920. doi:10.1088/0004-637X/693/2/1920

it's only around 2 billion years of age: Bechter E.B., Crepp J.R., Ngo H., Knutson H.A., Batygin K., Hinkley S., Muirhead P.S., et al., 2014, ApJ, 788, 2. doi:10.1088/0004-637X/788/1/2

Mercury orbits our own Sun at 0.4 au: https://solarsystem.nasa.gov/planets/mercury/in-depth/

more than ten times closer in at 0.03 au: Hebb L., Collier-Cameron A., Loeillet B., Pollacco D., Hébrard G., Street R.A., Bouchy F., et al., 2009, ApJ, 693, 1920. doi:10.1088/0004-637X/693/2/1920

the size, mass, and orbit of WASP-12b from previous observations: Hebb L., Collier-Cameron A., Loeillet B., Pollacco D., Hébrard G., Street R.A., Bouchy F., et al., 2009, ApJ, 693, 1920. doi:10.1088/0004-637X/693/2/1920

The answer: almost none: https://hubblesite.org/contents/news-releases/2017/news-2017-38.html

we would have concluded that it was a very blue planet: https://www.physicsclassroom.com/class/light/Lesson-2/Light-Absorption,-Reflection,-and-Transmission

absorbs 99.995% of all incoming light: https://news.mit.edu/2019/blackest-black-material-cnt-0913

as dark as freshly laid asphalt: https://web.archive.org/web/20070829153207/http://eetd.lbl.gov/HeatIsland/Pavements/Albedo/

or the open ocean: https://nsidc.org/cryosphere/seaice/processes/albedo.html

flecks of dust in the atmosphere would absorb light: Sing D.K., Lecavelier des Etangs A., Fortney J.J., Burrows A.S., Pont F., Wakeford H.R., Ballester G.E., et al., 2013, MNRAS, 436, 2956. doi:10.1093/mnras/stt1782

it's being shredded: https://www.nasa.gov/mission_pages/hubble/science/planet-eater.html

The planet is notably elongated: Li S.-L., Miller N., Lin D.N.C., Fortney J.J., 2010, Nature, 463, 1054. doi:10.1038/nature08715

more strongly pulled to its star than it is to itself: Li S.-L., Miller N., Lin D.N.C., Fortney J.J., 2010, Nature, 463, 1054. doi:10.1038/nature08715

the orbit of the planet is thought to be decaying: Yee S.W., Winn J.N., Knutson H.A., Patra K.C., Vissapragada S., Zhang M.M., Holman M.J., et al., 2020, ApJL, 888, L5. doi:10.3847/2041-8213/ab5c16

is expected to last only another 3 million years: Yee S.W., Winn J.N., Knutson H.A., Patra KC., Vissapragada S., Zhang M.M., Holman M.J., et al., 2020, ApJL, 888, L5. doi:10.3847/2041-8213/ab5c16

THE MOON SMELLS OF GUNPOWDER

the Apollo 16 astronauts Charles Duke and John Young: https://airandspace.si.edu/explore-and-learn/topics/apollo/apollo-program/landing-missions/apollo16.cfm

Really, really a strong odor to it: https://www.hq.nasa.gov/alsj/a16/a16.eva2post.html

Young: Yeah: https://www.hq.nasa.gov/alsj/a16/a16.eva1post.html

after returning from moonwalks: https://history.nasa.gov/alsj/TM-2005-213610.pdf

the only one to carry a geologist to the Moon: https://airandspace.si.edu/explore-and-learn/topics/apollo/apollo-program/landing-missions/apollo17-crew.cfm

Oh, it does, doesn't it?: https://www.hq.nasa.gov/alsj/a17/a17.eva1post.html

in the LRL [Lunar Receiving Lab] just to find out: https://www.hq.nasa.gov/alsj/a17/a17.eva1post.html

they wound up being contaminated with oxygen: https://science.nasa.gov/science-news/science-at-nasa/2006/30jan_smellofmoondust

several tumbles over the course of their moonwalks: https://www.youtube.com/watch?v=tqyK8v_f4Kg

ways to keep the dust from clinging so severely: https://www.nasa.gov/feature/goddard/2019/nasa-s-coating-technology-could-help-resolve-lunar-dust-challenge

problems similar to very severe pneumonia: https://science.nasa.gov/science-news/science-at-nasa/2005/22apr_dontinhale

an earthbound parallel called silicosis: https://www.lung.org/lung-health-diseases/lung-disease-lookup/silicosis

2,512 people in the US died of silicosis: https://www.ncbi.nlm.nih.gov/pmc/articles/PMC7315788/

the soil was not incredibly acidic or basic: https://cen.acs.org/articles/86/web/2008/06/Mars-Soil-pH-Measured.html

rocks present there *do* react strongly with water: Hurowitz J.A., Tosca N.J., McLennan S.M., Schoonen M.A.A., 2007, E&PSL, 255, 41. doi:10.1016/j.epsl.2006.12.004

'potentially creat[e] more deleterious effects': Pohlen, M., Carroll, D., Prisk, G.K. et al. Overview of lunar dust toxicity risk. npj Microgravity 8, 55 (2022). https://doi.org/10.1038/s41526-022-00244-1

YOU COULD GROW TURNIPS ON MARS SOIL IF IT WEREN'T FULL OF ROCKET FUEL

Phoenix Mars lander arrived at the surface of Mars in 2008: https://www.nasa.gov/mission_pages/phoenix/main/index.html

The Martian soil where Phoenix landed was between 8 and 9: https://www.nasa.gov/mission_pages/phoenix/news/phoenix-20080626.html

This is somewhere between 'seawater' and 'baking soda': https://www.usgs.gov/special-topic/water-science-school/science/ph-and-water?qt-science_center_objects=0#qt-science_center_objects

not all plants can grow in slightly basic soil: https://extension.umn.edu/vegetables/growing-potatoes#soil-testing-and-fertilizer-292660

'Martian Soil Could Grow Asparagus': https://www.discovermagazine.com/the-sciences/martian-soil-could-grow-asparagus; https://www.space.com/mars-soil-helps-asparagus-beans-iac-2019.html

'Martian soil could grow turnips, Phoenix finds': https://www.newscientist.com/article/dn14217-martian-soil-could-grow-turnips-phoenix-finds/

acidic soil, so strawberries were out: https://www.nasa.gov/mission_pages/phoenix/news/phoenix-20080626.html

leafy greens (lettuce and kale) to grow in greenhouses on Earth: https://skyandtelescope.org/astronomy-news/some-plants-grow-well-in-martian-soil/; https://eos.org/articles/tests-indicate-which-edible-plants-could-thrive-on-mars

another compound shortly afterwards: perchlorate: Hecht M.H., Kounaves S.P., Quinn R.C., West S.J., Young S.M.M., Ming D.W., Catling D.C., et al., 2009, Sci, 325, 64. doi:10.1126/science.1172466

this is what was in the solid rocket boosters: https://science.ksc.nasa.gov/shuttle/technology/sts-newsref/srb.html

0.5% of perchlorate salts: https://www.space.com/21554-mars-toxic-perchlorate-chemicals.html

likely calcium and magnesium perchlorate: Kounaves S.P., Chaniotakis N.A., Chevrier V.F., Carrier B.L., Folds K.E., Hansen V.M., McElhoney K.M., et al., 2014, Icar, 232, 226. doi:10.1016/j.icarus.2014.01.016

The Curiosity rover found perchlorate: Glavin D.P., Freissinet C., Miller K.E., Eigenbrode J.L., Brunner A.E., Buch A., Sutter B., et al., 2013, JGRE, 118, 1955. doi:10.1002/jgre.20144; Leshin L.A., Mahaffy P.R., Webster C.R., Cabane M., Coll P., Conrad P.G., Archer P.D., et al., 2013, Sci, 341, 1238937. doi:10.1126/science.1238937

and it has also been found *from orbit*: Keller J.M., Boynton W.V., Karunatillake S., Baker V.R., Dohm J.M., Evans L.G., Finch M.J., et al., 2006, JGRE, 111, E03S08. doi:10.1029/2006JE002679

you can kill bacteria *ten times faster*: Wadsworth, J., Cockell, C.S., 2017, *Sci Rep*, 7, 4662. doi.org/10.1038/s41598-017-04910-3

'a toxic cocktail of oxidants, iron oxides, perchlorates and UV irradiation': Wadsworth, J., Cockell, C.S., 2017, *Sci Rep*, 7, 4662. doi.org/10.1038/s41598-017-04910-3

1,000 to 10,000 times higher than what we get on Earth: Davila, A., Willson, D., Coates, J., McKay, C., 2013, *International Journal of Astrobiology*, 12(4), 321–5. doi:10.1017/473550413000189

more than the recommended dose of perchlorate: Davila, A., Willson, D., Coates, J., McKay, C., 2013, *International Journal of Astrobiology*, 12(4), 321–5. doi:10.1017/S1473550413000189

it can degrade into *even nastier* compounds: Quinn, Richard C., et al. 'Perchlorate Radiolysis on Mars and the Origin of Martian Soil Reactivity.' *Astrobiology*, vol. 13, no. 6, 2013, pp. 515–20

ClO_2^-: https://pubchem.ncbi.nlm.nih.gov/compound/Chlorine-dioxide#section=Safety-and-Hazards

ClO^-: https://pubchem.ncbi.nlm.nih.gov/compound/Calcium-hypochlorite#section=Safety-and-Hazards

'skin burns, loss of consciousness and vomiting': Quinn, Richard C., et al. 'Perchlorate Radiolysis on Mars and the Origin of Martian Soil Reactivity.' *Astrobiology*, vol. 13, no. 6, 2013, pp. 515–20

you've made poison plants out of poison dirt: Ha W., Suarez D.L., Lesch S.M., 2011, *Environmental Science & Technology*, 45(21), 9363–71. doi:10.1021/es2010094

tends to accumulate particularly strongly in the leaves: He, H., Gao, H., Chen, G. et al., 2013, *Environ Sci Pollut Res*, 20, 7301–08. https://doi.org/10.1007/s11356-013-1744-4; Ha W., Suarez D.L., Lesch S.M., 2011, *Environmental Science & Technology*, 45(21), 9363–71

They die: Eichler A., Hadland N., Pickett D., Masaitis D., Handy D., Perez A., Batcheldor D., et al., 2021, Icar, 354, 114022. doi:10.1016/j.icarus.2020.114022

they summarized their results thusly: Eichler A., Hadland N., Pickett D., Masaitis D., Handy D., Perez A., Batcheldor D., et al., 2021, Icar, 354, 114022. doi:10.1016/j.icarus.2020.114022

THE MOON ONCE HAD LAVA LAKES AND FIRE FOUNTAINS

closer to the Earth by about one-third (135,000 km closer): Bills B.G., Ray R.D., 1999, GeoRL, 26, 3045. doi:10.1029/1999GL008348

difficult period called the Late Heavy Bombardment: Gomes R., Levison H.F., Tsiganis K., Morbidelli A., 2005, Nature, 435, 466. doi:10.1038/nature03676

about 4 billion years to 3.5 billion years ago: Bottke W.F., Norman M.D., 2017, AREPS, 45, 619. doi:10.1146/annurev-earth-063016-020131

six landings is not enough to get all the interesting rocks: Bottke W.F., Norman M.D., 2017, AREPS, 45, 619. doi:10.1146/annurev-earth-063016-020131

tiny glass beads *all over* the surface: Wilson L., Head J.W., 2003, GeoRL, 30, 1605. doi:10.1029/2002GL016082

nearly the same mechanism of a shaken soda can: https://www.scientific american.com/article/why-does-a-shaken-soda-fi/

it only took 4.6 hours to rise from a depth of 500 km: Wilson L., Head J.W., 2017, Icar, 283, 146. doi:10.1016/j.icarus.2015.12.039

the surrounding 5–10 km of lunar surface: Wilson L., Head J.W., 2017, Icar, 283, 146. doi:10.1016/j.icarus.2015.12.039

between 10,000 and a million cubic meters of lava every second: Wilson L., Head J.W., 2017, Icar, 283, 146. doi:10.1016/j.icarus.2015.12.039

the inner layers could stay so hot: Wilson L., Head J.W., 2017, Icar, 283, 146. doi:10.1016/j.icarus.2015.12.039

Theoretical models: Wilson L., Head J.W., 2017, Icar, 283, 146. doi:10.1016/j.icarus.2015.12.039

exploding gas keeping the erupting material very hot: Wilson L., Head J.W., 2017, Icar, 283, 146. doi:10.1016/j.icarus.2015.12.039

Most eruptions were small overall: Wilson L., Head J.W., 2017, Icar, 283, 146. doi:10.1016/j.icarus.2015.12.039

A 1,200 km-long lava flow: Wilson L., Head J.W., 2017, Icar, 283, 146. doi:10.1016/j.icarus.2015.12.039

plausibly have taken a little under three days: Head J.W., Wilson L., Deutsch A.N., Rutherford M.J., Saal A.E., 2020, GeoRL, 47, e89509. doi:10.1029/2020GL089509

could keep going for nearly four months: Hulme G., 1973, ModGe, 4, 107

10 million cubic kilometers of lava: Head J.W., Wilson L., 1992, GeCoA, 56, 2155. doi:10.1016/0016-7037(92)90183-J

around 1.5 km in depth: Evans A.J., Soderblom J.M., Andrews-Hanna J.C., Solomon S.C., Zuber M.T., 2016, GeoRL, 43, 2445. doi:10.1002/2015GL067394; Williams K.K., Zuber M.T., 1998, Icar, 131, 107. doi:10.1006/icar.1997.5856

only laid down tens to hundreds of meters of lava: Weider S.Z., Crawford I.A., Joy K.H., 2010, Icar, 209, 323. doi:10.1016/j.icarus.2010.05.010

Probably 30,000–100,000 of them: Head J.W., Wilson L., Deutsch A.N., Rutherford M.J., Saal A.E., 2020, GeoRL, 47, e89509. doi:10.1029/2020GL089509

a lot of time between each eruption: Head J.W., Wilson L., Deutsch A.N., Rutherford M.J., Saal A. E., 2020, GeoRL, 47, e89509. doi:10.1029/2020GL089509

most prolific about 3 billion years ago: Head J.W., 1976, RvGSP, 14, 265. doi:10.1029/RG014i002p00265

probably occurred every million years or so: Head J.W., Wilson L., 1992, GeCoA, 56, 2155. doi:10.1016/0016-7037(92)90183-J

a year or two for the lava to cool completely: Head J.W., Wilson L., Deutsch A.N., Rutherford M.J., Saal A.E., 2020, GeoRL, 47, e89509. doi:10.1029/2020GL089509

as recently as 100 million to 50 million years ago: https://www.sciencemag.org/news/2014/10/recent-volcanic-eruptions-moon

SATURN'S LESS DENSE THAN WATER

Saturn is the second most massive planet: https://nssdc.gsfc.nasa.gov/planetary/factsheet/

known as differentiation: https://astronomy.swin.edu.au/cosmos/d/Differentiated+Object

a handy online chart listing the densities of all the planets: https://nssdc.gsfc.nasa.gov/planetary/factsheet/

Saturn is the only planet under this threshold: https://coolcosmos.ipac.caltech.edu/ask/113-Can-Saturn-really-float-on-water-

cans of diet soda: https://www.physics.upenn.edu/demolab/manumech/ms9.html

does not mean it would float in a bathtub: https://www.wired.com/2013/07/no-saturn-wouldnt-float-in-water/

if you could magic a planetary-sized bathtub into existence: https://www.nasa.gov/audience/forstudents/k-4/home/F_Saturn_Fun_Facts_K-4.html

VENUS'S SURFACE IS NEW

the 50 km-thick cloud layer: https://www.britannica.com/place/Venus-planet/The-atmosphere

sulfuric acid rain: Young A.T., 1973, Icar, 18, 564. doi:10.1016/0019-1035(73)90059-6

game of dodgeball that gravity was playing is still to be determined: https://www.nature.com/articles/d41586-018-01074-6

no *small* craters (smaller than 3 km across): https://www.psi.edu/epo/faq/venus.html

The craters which are there are really crisp: Schaber G.G., Strom R.G., Moore H.J., Soderblom L.A., Kirk R.L., Chadwick D.J., Dawson D.D., et al., 1992, JGR, 97, 13257. doi:10.1029/92JE01246

We basically have two options: Head J.W., Crumpler L.S., Aubele J.C., Guest J.E., Saunders R.S., 1992, JGR, 97, 13153. doi:10.1029/92JE01273

we do this 1) all at once: Schaber G.G., Strom R.G., Moore H.J., Soderblom L.A., Kirk R.L., Chadwick D.J., Dawson D.D., et al., 1992, JGR, 97, 13257. doi:10.1029/92JE01246

or 2) over time, randomly: Phillips R.J., Raubertas R.F., Arvidson R.E., Sarkar I.C., Herrick R.R., Izenberg N., Grimm R.E., 1992, JGR, 97, 15923. doi:10.1029/92JE01696

about 400 million years ago (give or take a few million years): Strom R.G., Schaber G.G., Dawsow D.D., 1994, JGR, 99, 10899. doi:10.1029/94JE00388

an affront to God in terms of the amount of lava that comes out per unit time: https://www.scientificamerican.com/article/model-suggests-toxic-transformation-on-venus/

the answer to this 'which' question is 'yes': Uppalapati S., Rolf T., Crameri F., Werner S.C., 2020, JGRE, 125, e06258. doi:10.1029/2019JE006258

were made by Magellan: https://solarsystem.nasa.gov/missions/magellan/in-depth/

the best global maps of Venus that we have: https://www.planetary.org/articles/the-venus-controversy

finally have new missions scheduled: https://www.nasa.gov/press-release/nasa-selects-2-missions-to-study-lost-habitable-world-of-venus

THE MOON'S WET

The Moon's wet: https://www.nasa.gov/press-release/nasa-s-sofia-discovers-water-on-sunlit-surface-of-moon; Honniball C.I., Lucey P.G., Li S., Shenoy S., Orlando T.M., Hibbitts C.A., Hurley D.M., et al., 2021, NatAs, 5, 121. doi:10.1038/s41550-020-01222-x

an Indian spacecraft, Chandrayaan-1: https://www.theguardian.com/world/2009/sep/24/water-moon-space-exploration-india

detected the signature of water from orbit: Pieters C.M., Goswami J.N., Clark R.N., Annadurai M., Boardman J., Buratti B., Combe J.-P., et al., 2009, Sci, 326, 568. doi:10.1126/science.1178658

LCROSS smashed itself into the Moon: https://www.nasa.gov/mission_pages/LCROSS/main/prelim_water_results.html

detected the signature of water away from the poles of the Moon: Honniball C.I., Lucey P.G., Li S., Shenoy S., Orlando T.M., Hibbitts C.A., Hurley D.M., et al., 2021, NatAs, 5, 121. doi:10.1038/s41550-020-01222-x

cold traps of permanent shade only centimeters across: Honniball C.I., Lucey P.G., Li S., Shenoy S., Orlando T.M., Hibbitts C.A., Hurley D.M., et al., 2021, NatAs, 5, 121. doi:10.1038/s41550-020-01222-x; Hayne P.O., Aharonson O., Schörghofer N., 2021, NatAs, 5, 169. doi:10.1038/s41550-020-1198-9\

allow more water to stay on the surface of the Moon: https://www.nasa.gov/feature/jpl/nasa-study-highlights-importance-of-surface-shadows-in-moon-water-puzzle

Sahara desert is still 100 times wetter than the Moon: https://www.nasa.gov/press-release/nasa-s-sofia-discovers-water-on-sunlit-surface-of-moon

volume of water at the poles (particularly the south pole) should be much larger: Hayne P.O., Aharonson O., Schörghofer N., 2021, NatAs, 5, 169. doi:10.1038/s41550-020-1198-9\

as dense as the Earth's atmosphere is where the ISS orbits: https://www.nasa.gov/mission_pages/LADEE/news/lunar-atmosphere.html

SOME OF TITAN'S LAKES MIGHT BE THE FLOODED REMAINS OF EXPLOSIONS

Some of Titan's lakes might be the flooded remains of explosions: Mitri G., Lunine J.I., Mastrogiuseppe M., Poggiali V., 2019, NatGe, 12, 791. doi:10.1038/s41561-019-0429-0

the only place in the solar system, aside from Earth, with lakes: Stofan E.R., Elachi C., Lunine J.I., Lorenz R.D., Stiles B., Mitchell K.L., Ostro S., et al., 2007, Nature, 445, 61. doi:10.1038/nature05438

it catches fire: https://pubchem.ncbi.nlm.nih.gov/compound/Methane

so far from the Sun and so cold (−179 C): https://www.jhuapl.edu/PressRelease/190627b

irregular, blob-shaped, *very* dark regions on the surface: Stofan E.R., Elachi C., Lunine J.I., Lorenz R.D., Stiles B., Mitchell K.L., Ostro S., et al., 2007, Nature, 445, 61. doi:10.1038/nature05438

a raised ridge, the most extreme rising 300 meters tall: Mitri G., Lunine J.I., Mastrogiuseppe M., Poggiali V., 2019, NatGe, 12, 791. doi:10.1038/s41561-019-0429-0

More typically, the ridges were 'only' 100 meters high: Birch S.P.D., Hayes A.G., Poggiali V., Hofgartner J.D., Lunine J.I., Malaska M.J., Wall S., et al., 2019, GeoRL, 46, 5846. doi:10.1029/2018GL078099

small lakes seem to have fortressed walls, somehow: Birch S.P.D., Hayes A.G., Poggiali V., Hofgartner J.D., Lunine J.I., Malaska M.J., Wall S., et al., 2019, GeoRL, 46, 5846. doi:10.1029/2018GL078099

craters tend to be round, and most of these ones aren't: Mitri G., Lunine J.I., Mastrogiuseppe M., Poggiali V., 2019, NatGe, 12, 791. doi:10.1038/s41561-019-0429-0

what's known as a maar: https://volcanoes.usgs.gov/vsc/glossary/maar.html

nitrogen is just going to catastrophically burst out of the ground: Tuttle Keane, J. 2019, *Nat Geosci*, 12, 789. https://doi.org/10.1038/s41561-019-0438-z

fill it with the methane rain: Tokano, T., McKay, C., Neubauer, F. et al., 2006, Nature, 442, 432–5. https://doi.org/10.1038/nature04948

methane rain that we've observed: Tomasko M.G., Archinal B., Becker T., Bézard B., Bushroe M., Combes M., Cook D., et al., 2005, Nature, 438, 765. doi:10.1038/nature04126

methane rain that we've observed falling on Titan: https://photojournal.jpl.nasa.gov/catalog/PIA12818

or perhaps a methane storm: Hueso R., Sánchez-Lavega A., 2006, Nature, 442, 428. doi:10.1038/nature04933

The spacecraft is called Dragonfly: https://dragonfly.jhuapl.edu

set to launch in 2026 and land in 2034: https://www.jhuapl.edu/Press Release/190627b

PLUTO'S SURFACE IS YOUNG, SOMEHOW

there are no craters bigger than 625 meters: Trilling D.E., 2016, PLoSO, 11, e0147386. doi:10.1371/journal.pone.0147386

bigger than this second limit, at 1,300 meters: https://earthobservatory. nasa.gov/images/1167/barringer-meteor-crater-arizona

only 50,000 years old at most: https://www.britannica.com/place/Meteor-Crater

between 40 craters and 50,000 craters larger than 30 km on the surface: Greenstreet S., Gladman B., McKinnon W.B., 2015, Icar, 258, 267. doi:10.1016/j.icarus.2015.05.026

nitrogen could just evaporate away into Pluto's atmosphere: Stern S.A., Porter S., Zangari A., 2015, Icar, 250, 287. doi:10.1016/j.icarus.2014. 12.006

some craters might also just slump flat: Stern S.A., Porter S., Zangari A., 2015, Icar, 250, 287. doi:10.1016/j.icarus.2014.12.006

But instead there were none:: Moore J.M., McKinnon W.B., Spencer J.R., Howard A.D., Schenk P.M., Beyer R.A., Nimmo F., et al., 2016, Sci, 351, 1284. doi:10.1126/science.aad7055

likely to be no older than 10 million years: Trilling D.E., 2016, PLoSO, 11, e0147386. doi:10.1371/journal.pone.0147386

This area of Pluto is indeed full of craters: Moore J.M., McKinnon W.B., Spencer J.R., Howard A.D., Schenk P.M., Beyer R.A., Nimmo F., et al., 2016, Sci, 351, 1284. doi:10.1126/science.aad7055

about a thousand craters have now been mapped: https://photojournal. jpl.nasa.gov/catalog/PIA20154

It begins to boil at −196 C (77 Kelvin): https://cryo.gsfc.nasa.gov/ introduction/liquid_helium.html

an even more frostbitten −210 C (63 K): https://cryo.gsfc.nasa.gov/ introduction/liquid_helium.html

from 10 km thick: Trowbridge A.J., Melosh H.J., Steckloff J.K., Freed A.M., 2016, Nature, 534, 79. doi:10.1038/nature18016

perhaps only 4 km deep: McKinnon W.B., Nimmo F., Wong T., Schenk P.M., White O.L., Roberts J.H., Moore J.M., et al., 2016, Nature, 534, 82. doi:10.1038/nature18289

filling in an ancient impact crater that we can no longer see: Schenk P.M., McKinnon W., Moore J., Nimmo F., Stern S.A., Weaver H., Ennico K., et al., 2015, DPS. https://ui.adsabs.harvard.edu/abs/2015DPS....4720006S/abstract

it can't be more than a million years old: Trowbridge A.J., Melosh H.J., Steckloff J.K., Freed A.M., 2016, Nature, 534, 79. doi:10.1038/nature18016

perhaps only 180,000 years old: Buhler P.B., Ingersoll A.P., 2017, LPI. bibcode: 2017LPI....48.1746B

The coelacanth is 400 times older than Pluto's surface: https://www.ncbi.nlm.nih.gov/pmc/articles/PMC1686207/

it also has glaciers: Howard A.D., Moore J.M., Umurhan O.M., White O.L., Anderson R.S., McKinnon W.B., Spencer J.R., et al., 2017, Icar, 287, 287. doi:10.1016/j.icarus.2016.07.006

SOME ASTEROIDS ARE JUST PILES OF RUBBLE IN SPACE

I'm pretty sure that has a 0% survival rate: https://www.nature.com/articles/s41467-020-15269-x

a ten-mile-wide asteroid: Collins G.S., Patel N., Davison T.M., Rae A.S.P., Morgan J.V., Gulick S.P.S., IODP-ICDP Expedition 364 Science Party, et al., 2020, NatCo, 11, 1480. doi:10.1038/s41467-020-15269-x

to the Yucatán peninsula: Goderis S., Sato H., Ferrière L., Schmitz B., Burney D., Kaskes P., Vellekoop J., et al., 2021, SciA, 7, eabe3647. doi:10.1126/sciadv.abe3647

Twelve in total, as of 2020: https://nssdc.gsfc.nasa.gov/planetary/planets/ asteroidpage.html

There have been two Hayabusa missions: https://nssdc.gsfc.nasa.gov/nmc/ spacecraft/display.action?id=2003-019A; https://nssdc.gsfc.nasa.gov/nmc/ spacecraft/display.action?id=2014-076A

'macroscopic particles (rubble) held together by their self gravity': Walsh K.J., 2018, ARA&A, 56, 593. doi:10.1146/annurev-astro-081817-052013

a density of around 2.5 grams per cubic centimeter: https://www.eoas. ubc.ca/ubcgif/iag/foundations/properties/density.htm

only about 1.2 grams per cubic centimeter: https://www.dlr.de/content/ en/articles/news/2019/03/20190822_the-near-earth-asteroid-ryugu- a-fragile-cosmic-rubble-pile.html

the rest is just bubbly rocks: Grott M., Biele J., Michel P., Sugita S., Schröder S., Sakatani N., Neumann W., et al., 2020, JGRE, 125, e06519. doi:10.1029/2020JE006519

rubble piles are the most common format: Walsh K.J., 2018, ARA&A, 56, 593. doi:10.1146/annurev-astro-081817-052013

a safe place for OSIRIS-REx to do its sample acquisition: https://www. nasa.gov/press-release/x-marks-the-spot-nasa-selects-site-for-asteroid- sample-collection

Bennu spins on its own axis once every 4.3 hours: https://en.wikipedia. org/wiki/101955_Bennu

Ryugu only a little bit slower at once every 7.6 hours: https://en.wikipedia. org/wiki/162173_Ryugu

these are not considered particularly rapid rotators: Walsh K.J., 2018, ARA&A, 56, 593. doi:10.1146/annurev-astro-081817-052013

Over time, this could crumble boulders into dust: Walsh K.J., 2018, ARA&A, 56, 593. doi:10.1146/annurev-astro-081817-052013

rotates more sedately, at once every twelve hours: https://en.wikipedia.org/wiki/25143_Itokawa

This type of object is called a 'contact binary': https://en.wikipedia.org/wiki/Contact_binary_(small_Solar_System_body)

14% of near-Earth asteroids are thought to be contact binaries: Taylor P. A., Howell E. S., Nolan M. C., Thane A. A., 2012, DPS. https://ui.adsabs.harvard.edu/abs/2012DPS....4430207T/abstract

JUPITER'S MAGNETIC FIELD WILL SHORT-CIRCUIT YOUR SPACECRAFT, BUT VENUS WILL JUST MELT IT

weasels chewing on power cables will both shut down the LHC: https://www.bbc.com/news/world-europe-36173247

'ALARA' – 'as low as reasonably achievable': https://www.cdc.gov/nceh/radiation/alara.html

rather than across the spacecraft (which fries it): https://www.nasa.gov/offices/nesc/articles/understanding-the-potential-dangers-of-spacecraft-charging

just down by about a factor of 800: https://www.coloradospacenews.com/junos-armor/

make it through its extended mission to 2025: https://www.jpl.nasa.gov/news/nasas-juno-mission-expands-into-the-future

The three principles of radiation avoidance: https://www.nrc.gov/about-nrc/radiation/protects-you/protection-principles.html

Venera 4 (1967): https://en.wikipedia.org/wiki/Venera_4

Venera 5 (1969): https://en.wikipedia.org/wiki/Venera_5

Venera 6 (1969): https://en.wikipedia.org/wiki/Venera_6

Venera 7 (1970): https://en.wikipedia.org/wiki/Venera_7

Venera 8 (1972): https://en.wikipedia.org/wiki/Venera_8

Venera 9 (1975): https://en.wikipedia.org/wiki/Venera_9

Venera 10 (1975): https://en.wikipedia.org/wiki/Venera_10

Venera 11 (1978): https://en.wikipedia.org/wiki/Venera_11

Venera 12 (1978): https://en.wikipedia.org/wiki/Venera_12

Venera 13 (1981): https://en.wikipedia.org/wiki/Venera_13

Venera 14 (1981): https://en.wikipedia.org/wiki/Venera_14

Vega 1 (1985): https://web.archive.org/web/20150520225951/http://solarsystem.nasa.gov/missions/profile.cfm?Sort=Target&Target=Venus&MCode=Vega_01

Vega 2 (1985): https://web.archive.org/web/20140414120546/http://solarsystem.nasa.gov/missions/profile.cfm?MCode=Vega_02&Display=ReadMore

EUROPA MIGHT GLOW IN THE DARK

a good mirror can reflect more than 95% of the light: https://en.wikipedia.org/wiki/Mirror#By_reflective_material

the Moon ... bounces back a flimsy 16% of the sunlight: https://astronomy.swin.edu.au/cosmos/a/Albedo

This category includes jellyfish: https://www.bbc.com/news/science-environment-14882008; https://www.theverge.com/2013/12/30/5256732/scientists-create-glow-in-the-dark-pigs-using-jellyfish-dna; https://www.technologyreview.com/2009/05/27/124843/glowing-monkeys-inherit-jellyfish-genes/

emeralds: http://www.geo.utexas.edu/courses/347k/redesign/Gem_Notes/Beryl/beryl_triple_frame.htm

rubies: https://www.gemsociety.org/article/natural-rubies-fluoresce/

scorpions: Lawrence, R.F., 1954, Fluorescence in Arthropoda. *Journal of the Entomological Society of South Africa*, 17: 167–70; Herbert L. Stahnke, UV Light, A Useful Field Tool, 1972, *BioScience*, Volume 22, Issue 10, 604–07, https://doi.org/10.2307/1296207

not a great thing to be employed to perform: https://www.cnn.com/style/article/radium-girls-radioactive-paint/index.html

predictions from a laboratory on Earth: Gudipati M.S., Henderson B.L., Bateman F.B., 2021, NatAs, 5, 276. doi:10.1038/s41550-020-01248-1

and not found anything: Sparks W.B., McGrath M., Hand K., Ford H.C., Geissler P., Hough J.H., Turner E.L., et al., 2010, IJAsB, 9, 265. doi:10.1017/S1473550410000285

say, for example, Europa: Cooper J.F., Johnson R.E., Mauk B.H., Garrett H.B., Gehrels N., 2001, Icar, 149, 133. doi:10.1006/icar.2000.6498

strike ice with radiation like this, it will glow: Ghormley J.A., 1956, JChPh, 24, 1111. doi:10.1063/1.1742702; Grossweiner L.I., Matheson M.S., 1954, JChPh, 22, 1514. doi:10.1063/1.1740451

A glow happened: Gudipati M.S., Henderson B.L., Bateman F.B., 2021, NatAs, 5, 276. doi:10.1038/s41550-020-01248-1

color code of #4AFF00, if you want to check yourself: https://academo.org/demos/wavelength-to-colour-relationship/

The Europa Clipper: https://www.jpl.nasa.gov/missions/europa-clipper

should be carrying a visible light camera: https://europa.nasa.gov/mission/about/

SATURN'S RINGS ARE FALLING APART

Saturn's main features vs. the Earth, courtesy of NASA: https://nssdc. gsfc.nasa.gov/planetary/factsheet/saturnfact.html

Somewhere between 95% and 99%: https://solarsystem.nasa.gov/ missions/cassini/science/rings/;https://science.sciencemag.org/content/362/ 6410/eaat3185

At their thickest, they're about a kilometer tall: Brahic A., Sicardy B., 1981, Nature, 289, 447. doi:10.1038/289447a0

more typically only about 10 meters: https://solarsystem.nasa.gov/planets/ saturn/in-depth/#otp_rings

The rings are made of trillions of small pieces of ice: https://solarsystem. nasa.gov/missions/cassini/science/rings/

to microscopically fine ice dust: https://www.nasa.gov/press-release/ goddard/2018/ring-rain

This works out to 1.5×10^{19} kg: https://solarsystem.nasa.gov/moons/ saturn-moons/mimas/by-the-numbers/

shredded a big comet: Dones L., 1991, Icar, 92, 194. doi:10.1016/0019-1035(91)90045-U

with a tremendously massive ring system: Crida A., Charnoz S., Hsu H.-W., Dones L., 2019, NatAs, 3, 967. doi:10.1038/s41550-019-0876-y

and they did, 22 times: https://solarsystem.nasa.gov/missions/cassini/ mission/grand-finale/grand-finale-orbit-guide/

the southernmost point of England to the northernmost point in Scotland is 1,000 km: https://www.britannica.com/place/United-Kingdom

To the tune of 432 to 2,870 kilograms of water every second: O'Donoghue J., Moore L., Connerney J., Melin H., Stallard T.S., Miller S., Baines K.H., 2019, Icar, 322, 251. doi:10.1016/j.icarus.2018.10.027

only another 300 million years: O'Donoghue J., Moore L., Connerney J., Melin H., Stallard T. S., Miller S., Baines K. H., 2019, Icar, 322, 251. doi:10.1016/j.icarus.2018.10.027

you can extrapolate your way to a much larger number: Mitchell D.G., Perry M.E., Hamilton D.C., Westlake J.H., Kollmann P., Smith H.T., Carbary J.F., et al., 2018, Sci, 362, aat2236. doi:10.1126/science.aat2236

about as fine as the soot in smoke: https://www.nasa.gov/feature/jpl/groundbreaking-science-emerges-from-ultra-close-orbits-of-saturn

$(1.4 \times 10^{24}$ of them), assuming random close packing: https://www.nature.com/articles/nature06981

methane, among other molecules: Waite J.H., Perryman R.S., Perry M.E., Miller K.E., Bell J., Cravens T.E., Glein C.R., et al., 2018, Sci, 362, aat2382. doi:10.1126/science.aat2382

could only provide for this kind of loss for another 7,000 years: Waite J.H., Perryman R.S., Perry M.E., Miller K.E., Bell J., Cravens T.E., Glein C.R., et al., 2018, Sci, 362, aat2382. doi:10.1126/science.aat2382

might be as young as 10 million years old: Iess L., Militzer B., Kaspi Y., Nicholson P., Durante D., Racioppa P., Anabtawi A., et al., 2019, Sci, 364, aat2965. doi:10.1126/science.aat2965

CERES ONCE HAD VOLCANOES THAT ERUPTED WITH SALT WATER

'This is truly unexpected and still a mystery to us': https://www.npr.org/sections/thetwo-way/2015/02/26/389245969/nasa-sees-bright-spots-on-dwarf-planet-in-our-solar-system

called hexahydrate magnesium sulfate: Nathues A., Hoffmann M., Schaefer M., Le Corre L., Reddy V., Platz T., Cloutis E.A., et al., 2015, Nature, 528, 237. doi:10.1038/nature15754

later research changed the salt to sodium carbonate: https://www.nasa.gov/feature/jpl/recent-hydrothermal-activity-may-explain-ceres-brightest-area

a pathway that leads the water to the surface: Nathues A., Schmedemann N., Thangjam G., Pasckert J.H., Mengel K., Castillo-Rogez J., Cloutis E.A., et al., 2020, NatAs, 4, 794. doi:10.1038/s41550-020-1146-8

a particularly young crater called Occator Crater: Nathues A., Hoffmann M., Schaefer M., Le Corre L., Reddy V., Platz T., Cloutis E.A., et al., 2015, Nature, 528, 237. doi:10.1038/nature15754

might still be leaking salt water *now*: https://www.jpl.nasa.gov/news/mystery-solved-bright-areas-on-ceres-come-from-salty-water-below

sides at 30–40 degree angles: Ruesch O., Platz T., Schenk P., McFadden L.A., Castillo-Rogez J.C., Quick L.C., Byrne S., et al., 2016, Sci, 353, aaf4286. doi:10.1126/science.aaf4286

it probably had company in the past: Sori M.M., Byrne S., Bland M.T., Bramson A.M., Ermakov A.I., Hamilton C.W., Otto K.A., et al., 2017, GeoRL, 44, 1243. doi:10.1002/2016GL072319

it should slump downwards: https://news.agu.org/press-release/new-research-shows-ceres-may-have-vanishing-ice-volcanoes/

22 more probable cryovolcanoes: Sori M.M., Sizemore H.G., Byrne S., Bramson A.M., Bland M.T., Stein N.T., Russell C.T., 2018, NatAs, 2, 946. doi:10.1038/s41550-018-0574-1

40 kilometers down and hundreds of kilometers across: https://www.jpl.nasa.gov/news/mystery-solved-bright-areas-on-ceres-come-from-salty-water-below

TRITON ORBITS BACKWARDS AND IS DOOMED

Neptune has fourteen known moons: https://nssdc.gsfc.nasa.gov/planetary/factsheet/neptunefact.html

demigods, deities, and water-related spirits: https://solarsystem.nasa.gov/
moons/neptune-moons/overview/

Earth's Moon is actually bigger than Triton: https://en.wikipedia.org/wiki/
List_of_natural_satellites

more likely that Triton was once a Pluto-like world: https://solarsystem.
nasa.gov/moons/neptune-moons/triton/in-depth/; McKinnon W.B., 1984,
Nature, 311, 355. doi:10.1038/311355a0

how exactly Triton could have been captured: Ćuk M., Gladman B.J.,
2005, ApJL, 626, L113. doi:10.1086/431743; Rufu R., Canup R.M.,
2017, AJ, 154, 208. doi:10.3847/1538-3881/aa9184

its companion might have been more massive than itself: Agnor C.B.,
Hamilton D.P., 2006, Nature, 441, 192. doi:10.1038/nature04792

**there are a bunch of other moons that your model can't destroy
or eject**: Holman M.J., Kavelaars J.J., Grav T., Gladman B.J., Fraser
W.C., Milisavljevic D., Nicholson P.D., et al., 2004, Nature, 430, 865.
doi:10.1038/nature02832

Triton was captured really early: Nogueira E., Brasser R., Gomes R.,
2011, Icar, 214, 113. doi:10.1016/j.icarus.2011.05.003; Li D., Christou
A.A., 2020, AJ, 159, 184. doi:10.3847/1538-3881/ab7cd5

lower than Jupiter's 79, Saturn's 82, or Uranus' 27 known moons: https://
nssdc.gsfc.nasa.gov/planetary/factsheet/jupiterfact.html; https://nssdc.gsfc.
nasa.gov/planetary/factsheet/saturnfact.html; https://nssdc.gsfc.nasa.gov/
planetary/factsheet/uranusfact.html

**flinging moons of Neptune out into ... the furthest reaches of the solar sys-
tem**: Rufu R., Canup R.M., 2017, AJ, 154, 208. doi:10.3847/1538-3881/
aa9184

crashing a bunch of other moons into each other: Ćuk M., Gladman B.J.,
2005, ApJL, 626, L113. doi:10.1086/431743

It's going to be ripped apart by Neptune: McCord T.B., 1966, AJ, 71, 585. doi:10.1086/109967

came up with a number of 1.5 billion years: Chyba C.F., Jankowski D.G., Nicholson P.D., 1989, A&A, 219, L23. https://ui.adsabs.harvard.edu/abs/1989A%26A...219L..23C/abstract

the language is 'probably unrealistically short': Chyba C.F., Jankowski D.G., Nicholson P.D., 1989, A&A, 219, L23. https://ui.adsabs.harvard.edu/abs/1989A%26A...219L..23C/abstract

Jillian Scudder's previous book,
Astroquizzical, **is available as both**
a paperback (with colour plates) and as a
large-format hardback, fully illustrated edition:

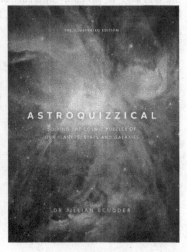

Paperback
ISBN: 978-178578-412-5

Illustrated hardback
ISBN: 978-178578-755-3
(USA/Canada: 978-026204-672-5)

How often do we think about the myriad links
between our home planet and the universe?
Astrophysicist and blogger Jillian Scudder takes us
on an enthralling journey to reveal our place in the
wider cosmos. From shooting stars, to colliding black
holes, to the fate of entire galaxies, *Astroquizzical*
is a stunning voyage through space and time.